Protein phosphorylation in control mechanisms.
    Edited by F. Huijing [and] E. Y. C. Lee.
    New York, Academic Press, 1973.
    xiv, 321 p. illus. (Miami winter symposia,
    v. 5)
    "Proceedings of the Miami winter symposia,
January 15-16, 1973, organized by the Dept.
of Biochemistry, University of Miami School
of Medicine, Miami, Florida."
    Includes bibliographical references.

                    (continued next card)

Protein Phosphorylation in Control Mechanisms

miami winter symposia

1. W. J. Whelan and J. Schultz, editors: HOMOLOGIES IN ENZYMES AND META-BOLIC PATHWAYS and METABOLIC ALTERATIONS IN CANCER,* 1970

2. D. W. Ribbons, J. F. Woessner, Jr., and J. Schultz, editors: NUCLEIC ACID-PROTEIN INTERACTIONS and NUCLEIC ACID SYNTHESIS IN VIRAL INFECTION,* 1971

3. J. F. Woessner, Jr. and F. Huijing, editors: THE MOLECULAR BASIS OF BIOLOGICAL TRANSPORT, 1972

4. J. Schultz and B. F. Cameron, editors: THE MOLECULAR BASIS OF ELECTRON TRANSPORT, 1972

5. F. Huijing and E. Y. C. Lee, editors: PROTEIN PHOSPHORYLATION IN CONTROL MECHANISMS, 1973

6. J. Schultz and H. G. Gratzner, editors: THE ROLE OF CYCLIC NUCLEOTIDES IN CARCINOGENESIS, 1973

*Published by North-Holland Publishing Company, Amsterdam, The Netherlands.

MIAMI WINTER SYMPOSIA — VOLUME 5

Protein Phosphorylation in Control Mechanisms

edited by

F. Huijing
E. Y. C. Lee

DEPARTMENT OF BIOCHEMISTRY
UNIVERSITY OF MIAMI SCHOOL OF MEDICINE
MIAMI, FLORIDA

*Proceedings of the Miami Winter Symposia, January 15-16, 1973,
organized by the Department of Biochemistry, University of Miami
School of Medicine, Miami, Florida*

ACADEMIC PRESS   New York and London 1973
A Subsidiary of Harcourt Brace Jovanovich, Publishers

ACADEMIC PRESS, INC.
111 Fifth Avenue, New York, New York 10003

United Kingdom Edition published by
ACADEMIC PRESS, INC. (LONDON) LTD.
24/28 Oval Road, London NW1

Library of Congress Cataloging in Publication Data

Main entry under title.

Protein phosphorylation in control mechanisms.

    (Miami winter symposia, v. 5)
    "Proceedings of the Miami winter symposia, January 15-
16, 1973, organized by the Dept. of Biochemistry,
University of Miami School of Medicine, Miami, Florida."
    1. Phosphorylation–Congresses. 2. Proteins–
Congresses. 3. Cellular control mechanisms–
Congresses. 4. Cyclic adenylic acid–Congresses.
I. Huijing, F., ed. II. Lee, Ernest Yee Chung, ed.
III. Miami, University of, Coral Gables, Fla. Dept. of
Biochemistry. IV. Series. [DNLM: 1. Phos-
phorylase kinase–Congresses. 2. Phosphotransferases–
Congresses. W3 M1202 v. 5 1973. XNLM: [QU 135 P967
1973]]
QP535.P1P76 1973   599'.01'9214   72-9329
ISBN 0–12–360950–X

CONTENTS

CONTENTS

CONTENTS

Free Communications

CONTENTS

SPEAKERS, CHAIRMEN, AND DISCUSSANTS

C. Abell, Department of Human Biological Chemistry and Genetics, University of Texas, Galveston, Texas

V. G. Allfrey, Department of Cell Biology, Rockefeller University, New York

S. H. Appel, Division of Neurology, Duke Hospital, Durham, North Carolina

J. Ashmore, Department of Pharmacology, Indiana University School of Medicine, Indianapolis, Indiana

S. A. Assaf, Department of Medicine, University of Miami School of Medicine, Miami, Florida

M. Bitensky, Yale University, New Haven, Connecticut

A. Braunwalder, Department of Microbiology, CIBA Geigy Corp., Summit, New Jersey

B. L. Brown, Institute of Nuclear Medicine, London, England

R. Chalkley, Department of Biochemistry, University of Iowa, Iowa City, Iowa

B. S. Cooperman, Chemistry Department, University of Pennsylvania, Philadelphia, Pennsylvania

J. D. Corbin, Department of Physiology, Vanderbilt University, Nashville, Tennessee

C. Dalton, Hoffman–La Roche Inc., Biochemical Pharmacology, Nutley, New Jersey

G. H. Dixon, (Session Chairman), Department of Biochemistry, University of Sussex, Salmer Brighton, England

J. Fessenden-Raden, Department of Biochemistry, Cornell University, Ithaca, New York

P. A. Galand, Nuclear Medicine, Free University of Brussels, Brussels, Belgium

G. N. Gill, Department of Medicine, University of California, San Diego, California

A. G. Gornall, Department of Clinical Biochemistry, University of Toronto, Toronto, Ontario, Canada

P. Greengard, (Session Chairman), Department of Pharmacology, Yale University Medical School, New Haven, Connecticut

L. R. Gurley, Los Alamos Scientific Laboratory, University of California, Los Alamos, California

B. L. Horecker, (Session Chairman), Department of Physiological Chemistry, Albert Einstein College of Medicine, New York

F. Huijing, Department of Biochemistry, University of Miami School of Medicine, Miami, Florida

E. M. Johnson, Rockefeller University, New York

G. A. Kimmich, Department Radiological Biology and Biophysics, University of Rochester, New York

E. G. Krebs, Department of Biological Chemistry, University of California, Davis, California

G. Krishna, National Institute of Health, Bethesda, Maryland

T. A. Langan, Department of Pharmacology, University of Colorado Medical School, Denver, Colorado

G. S. Levey, Department of Medicine, University of Miami School of Medicine, Miami, Florida

R. W. Longton, Naval Medical Research Institute, National Naval Medical Center, Bethesda, Maryland

H. R. Mahler, Department of Chemistry, Indiana University, Bloomington, Indiana

J. M. Marsh, Department of Biochemistry, University of Miami School of Medicine, Miami, Florida

O. J. Martelo, Department of Medicine, University of Miami School of Medicine, Miami, Florida

K. S. McCarty, Department of Biochemistry, Duke Medical Center, Durham, North Carolina

J. P. Miller, ICN Nucleic Acid Research Institute, Irvine, California

C. Moore, Albert Einstein College of Medicine, New York

C. R. Park, Department of Physiology, Vanderbilt University, Nashville, Tennessee

R. Piras, Instituto de Investigaciones, Bioquimicas Fundacion Campomar, Buenos Aires, Argentina

L. J. Reed, Clayton Foundation, University of Texas, Austin, Texas

E. Reimann, Department of Biochemistry, Medical College of Ohio, Toledo, Ohio

M. S. Rose, ICI Industrial Hygiene Research Laboratories, Cheshire, England

O. M. Rosen, Department of Medicine and Molecular Biology, Albert Einstein College of Medicine, New York

H. Rosenkranz, Director of Biochemistry, Mason Research Institute, Worcester, Massachusetts

J. Roth, Section of Biochemistry and Biophysics, University of Connecticut, Storrs, Connecticut

H. Segal, Biology Department, State University of New York, Buffalo, New York

R. Sharma, Department of Biochemistry, Veterans Administration Hospital, Memphis, Tennessee

K. Shelton, Department of Biochemistry Medical College of Virginia, Richmond, Virginia

H. Sheppard, Hoffman-La Roche Inc., Biochemical Pharmacology, Nutley, New Jersey

K. S. Sidhu, Endocrine Laboratory, University of Miami School of Medicine, Miami, Florida

T. Soderling, Department of Physiology, Vanderbilt University, Nashville, Tennessee

H. K. Stanford, (Session Chairman), President of the University of Miami, Coral Gables, Florida

D. Steinberg, Department of Medicine, University of California, San Diego, California

S. Strada, Program in Pharmacology, University of Texas Medical School, Houston, Texas

M. Tao, Department of Biological Chemistry, University of Illinois, Chicago, Illinois

V. Tomasi, Department of Physiology, University of Ferrara, Ferrara, Italy

W. J. Whelan, Chairman, Department of Biochemistry, University of Miami School of Medicine, Miami, Florida

W. D. Wicks, Department of Pharmacology, University of Colorado Medical Center, Denver, Colorado

J. Willis, P-L Biochemicals, Inc., Milwaukee, Wisconsin

C. Zeilig, Vanderbilt University, Nashville, Tennessee

PREFACE

In January, 1969, the Department of Biochemistry of the University of Miami School of Medicine and the University-affiliated Papanicolaou Cancer Research Institute joined in presenting two symposia on biochemical topics. These symposia have begun to develop as a tradition over the years and have attracted national interest.

In 1970, the two symposia were entitled "Homologies in Enzymes and Metabolic Pathways" and "Metabolic Alterations in Cancer." The full report of this meeting was published as the first volume of a continuing series under the title "Miami Winter Symposia."

In 1971, the value of the series was enhanced by including all discussions as well as the full text of the reports. In 1972 it was decided to publish the results of the two symposia separately, in order to prevent the volumes from becoming unwieldy and to permit a wider scope in the selection of the topics for the two symposia.

This volume, the fifth in the series, contains the proceedings of the Department of Biochemistry's January 1973 symposia entitled "Protein Phosphorylation in Control Mechanisms" and will appear simultaneously with volume 6, the proceedings of the Papanicolaou Cancer Research Institute symposium on "The Role of Cyclic Nucleotides in Carcinogenesis."

The success of the Miami Winter Symposia is indicated by a continued increase in the number of registrants, which this year reached 433. The organizers may have to consider putting an upper limit on the number of registrants in order to preserve the nature of the meeting.

An innovation this year was the presentation of short communications on the free day between the two symposia. The abstracts of the communications related to protein phosphorylation are printed in this volume.

Associated with the symposia is a featured lecture of autobiographical nature, named in honor of the Department of Biochemistry's distinguished Visiting Professor, Professor Feodor Lynen. Past Feodor Lynen Lecturers are Dr. George Wald, Dr. Arthur Kornberg, and Dr. Harland G. Wood. This year's Lynen Lecturer was Dr. Earl W. Sutherland, who unfortunately fell ill shortly before the symposia and was hospitalized in the University of Miami Medical Center during the symposia. We are very grateful to Dr. C. R. Park, longtime friend and colleague of Dr. Sutherland, who at short notice prepared and delivered a biographical lecture on behalf of Dr. Sutherland. This lecture forms the opening paper of this volume.

The symposia are organized to ensure publication as rapidly as possible. The editors want to thank all the contributors, the speakers for submitting their manuscripts promptly, the discussants for editing their discussions, and especially the secretarial staff.

We also acknowledge with gratitude the financial assistance of the Howard Hughes Medical Institute, the Departments of Anesthesiology and Dermatology of the University of Miami, Coulter Electronics, Inc., Eli Lilly and Company, MC/B Manufacturing Chemists, Smith Kline and French, and E. R. Squibb and Son.

In 1974 the topics of the Miami Winter Symposia will be "Carbohydrates and Cell-Surface Recognition" and "Membrane Transportations in Neoplasia." The Lynen Lecturer will be Dr. Luis F. Leloir. The symposia will be held from January 14-18, 1974.

F. Huijing
E. Y. C. Lee

THE FOURTH FEODOR LYNEN LECTURE:

My Life and Cyclic AMP

Earl W. Sutherland

Professor of Physiology, Vanderbilt University
School of Medicine, Nashville, Tennessee

with a preface and editing by

Charles R. Park

Professor and Chairman
Department of Physiology
Vanderbilt University School of Medicine
Nashville, Tennessee

Dr. Earl W. Sutherland was to give this talk but could
not do so because of sudden and serious illness. I had
misgivings when he asked me to speak for him, but my enthu-
siasm developed as I realized that here was an opportunity
to present the man and his work to you in an unusual way.
Earl Sutherland being very human, did not have his lecture
prepared two weeks ahead of time and consequently could not
give me a manuscript to read to you. This is perhaps fortunate. Sutherland is like his late neighbor, Harry Truman,
who learned during his brilliant campaign for the presidency
that he transmitted his character and special way of thinking much more effectively when he spoke without a text. To
be frank, Sutherland's talks are sometimes minor disasters,
but, when the occasion inspires him, his lectures are attractive, effective and quite unique. A few months ago, he
gave an account of his scientific work on an occasion at
Vanderbilt which provided this inspiration. I have taken
the present presentation from a tape of that talk, editing
for clarity but trying to preserve and transmit the

essential informality and flavor of my colleague's character.

A few weeks before this symposium, Sutherland told me
that he had turned down perhaps as many as 60 lecture
invitations within the past year. The invitation to speak
here in Miami was one of the very few he accepted. This
indicates clearly that he felt this occasion be a special
one. He was also anxious that I transmit to you the plea-
sure it gave him to contribute to the honor of his friend,
Feodor Lynen.

I will preface Sutherland's scientific account with
some biographical data and thoughts about him as a scien-
tist and as a man. Sutherland was born 58 years ago in
Burlingame, Kansas (pop. about 1,000), the fifth child in
a family of six. He grew up experiencing a number of years
of comfortable prosperity but then felt the pinch of auster-
ity when the depression brought hard times to his father's
dry goods business. One is tempted to ascribe prominent
features of Sutherland's nature to patterns of living
established in those early years. He was brought up to
value highly a strictly interpreted code of honesty in
human relationships. This code strongly influences his
interaction with his fellows, attracting him to the plain
dealers and turning him against flatterers and sycophants.
He has been consistently able to turn a dispassionate eye
on his own research and consequently his publications are
singularly free of the wish-fulfilling overevaluations in
which most of us indulge. This objectivity has played an
important part in the steady progression of his work and
its singular freedom from errors, a most unusual feature
even in the works of other top-level scientists.

What are the secrets of Sutherland's scientific
success? He has a remarkable ability to recall the count-
less experiments carried out by him and his associates over
the past 25 years. He spends many hours of thoughts and
discussion integrating these past experiences with newer
findings in order to lay out a systematic approach to a
problem. Sutherland is not one to plunge impulsively into
laboratory experimentation. He exemplifies Sir Frederick
Hopkins' formula for research endeavor:

"In a country rich in gold, observant wayfarers may

2

find occasional nuggets on their path, but only systematic mining can provide the currency of nations."

But a most important feature of Sutherland's success lies in his originality. While the vast majority of scientists plan their studies to follow lines of thoughts proposed by others in related fields, Sutherland may develop and act on quite new concepts. His discoveries of enzyme activation by enzymatic phosphorylation and of the intracellular second messenger system (cyclic adenylate) introduced concepts which were quite new to the scientific world.

Sutherland's youth in Kansas appears also to have given him a life-long love for the country, with its opportunities for hunting and fishing. You may smile with amusement, or envy, at his passion for fishing, but it would be hard to overestimate what it has meant to him. Fishing provides the release for a man whose relaxed exterior conceals powerful drives and tensions. Nobel prizes are not won by jokes and banter (at which, incidentally Sutherland is a specialist!). Sutherland has never been attracted, except as a visitor, to big city life. This is not to imply any lack of sophistication, because he is as thoroughly urbane and knowledgeable as you would expect a good observer to be who has traveled the world very widely. Nashville, his present home, is sufficiently countrified to suit him well; Vanderbilt provides men and tools by which his investigations can flourish, and the easily accessible, happily climated countryside of Tennessee gives him his release, in summer time, at least. Recently, Sutherland has spent much of his winter-time in the Florida Keys. He claims that some of his best thoughts have come on the waters of Tennessee and Florida. Some fish have come too.

When Sutherland began his talk at Vanderbilt, to which I alluded earlier, he directed some remarks to me which you who know him will recognize as being in character. He opined that he had learned something about fishing from me-to his surprize-how to appreciate catching small fish! He recalled a canoe trip. The canoe seemed to him remarkably small, hardly big enough to accomodate his torso and my feet, (which I add parenthetically are noted for their size). The fishing was exceptionally good; we caught

3

two bass that day! They were unusually large and how could they be accommodated in this crowded canoe? I proved my resourcefulness, however, by taking off a shoe. First, the 3 lb. bass was slipped into it, and then the 2 lb. bass and then a towel to hold them in place!

Earl then moved into the substance of his talk which, I remind you, was informal and without script. The pronoun, I, will now be Earl speaking. (1,2,3)

"I have been asked to discuss our research work to-day--and I really mean our rather than my work. Some of you will know my recent associates, Bill Butcher, Al Robison, and my present collaborators, Joel Hardman, Ren-jye Ho and Jim Davis. You know that these people are more active than I--and it isn't only a question of acti-vity, because these people are bright and are in there with ideas. I want also to mention young members of my group like Dr. Roger Johnson, Dr. Charles Sutherland, and recently, the medical student, Arthur Broadus. Now, as I go along, I won't have time to give credit to all people who have worked with me in this field, but I would like to mention especially Dr. Wosilait and Dr. Rall who were very important in the earlier days of this research work. As you know, after quite a long latent period, there has been a tremendous explosion of effort in the cyclic AMP world. I would estimate there might be 2,000 workers in the field now and many of the major drug companies are active. One of the heads of a commercial lab recently wrote me that his work was coming along very well and he hoped to have a couple of compounds coming out for clinical use. Well, having said these things, I would like now to show you some experiments illustrating the development of research in cAMP.

During the past several years a concept has developed about how some hormones act and this is summarized here (Fig. 1). About half the hormones work through the mechanism depicted here and roughly half the others are released by this cyclic nucleotide. So, of course, a lot of work on cAMP has been centered around how a hormone works. Now you have more hormones than you think. Most men figure they have at least one, and most women think likewise, but really there are lots of these floating

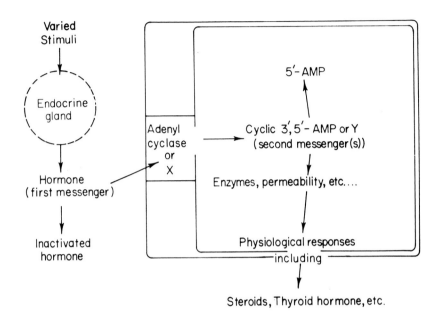

Fig. 1: The second messenger concept (from 3).

around. Even in your brain, you have neuro-hormones and
they have to do with thinking and memory. So perhaps we
could add up at least 28 agents in this category, but, as
I will mention later, the whole story of cAMP includes
more than just hormones. But, anyway, we started out in
this area and eventually developed the concept shown in
this figure which provides a general summary of the action
of approximately half the hormones. You have an endocrine
gland, and there are many stimuli that cause that gland to
secrete a hormone, which by definition from the Greek, is
a messenger, and we call it a first messenger. It generally enters into the blood stream and travels to a cell.
However, in some cases it doesn't travel through the blood
stream but is injected right onto the cell. For example,
this applies to all the sympathetic nerves in the body

5

which govern a lot of involuntary regulation such as blood
pressure and heart rate. Whether or not a hormone is
released directly on a cell or into the blood, you have
the possibility of inactivation or binding, so modulation
can occur at this point. Now, when the hormone contacts
the cells, it reacts with a particulate portion of the
cell, and, where this has been studied in detail, it is
the plasma membrane. You see, this is very convenient, the
first messenger can come to the plasma membrane on the outside, reacts with this system, and then causes an increase
or decrease in the second messenger, cyclic 3',5'-AMP, on
the inside of the cell. The cyclic 3',5'-AMP then goes
ahead to change enzymes and permeability and a great
number of other things including the secretion of other
hormones.

Now I want to return to the olden days. This experiment shown in Fig. 2 was done many years ago, so long that
it recalls many early events and things -- but I can't
talk about those here. Well, what we did was to choose
two hormones, adrenalin and glucagon, which both increase
the blood sugar. Because they work very rapidly, we
thought we might be able to trace down what they do. We
worked with living cells because the hormones didn't seem
to have any real effect on isolated enzymes. We found that
when we looked at the counts in inorganic phosphate that
went into the intermediates in glycogen breakdown in a
liver slice preparation that they were increased by either
hormone. And finally, over a several year period, this
and other studies led us to the conclusion that certain
enzymes had to be activated. Now why were we so slow -
it took us several years to get here? Part of the difficulty was working with intact cells but part of the slow-
ness was due to the fact that the textbooks told us that
the enzyme, phosphorylase, catalyzed the synthesis of
glycogen just as well or better than the breakdown. So
epinephrine or glucagon, which always breaks down glycogen,
couldn't be affecting this enzyme because synthesis did
not take place. Now, I do believe we should have a lot
of respect for words from our parents and teachers but
maybe not for all the things that are in the text books,
you see. Actually, I am being quite conservative in this
statement. Anyway, this held us up for quite a while, but
Leloir finally discovered there was another enzyme which

synthesized glycogen, and that there was so much inorganic
phosphate inside of our cells that phosphorylase could
operate only in the direction of glycogen breakdown.

INTERMEDIATE	CONTROL	GLUCAGON	EPINEPHRINE
	cts $\cdot$ mg $P^{-1} \cdot$ min$^{-1} \cdot$ g liver^{-1} (x 10^{-2})		
G-1-P	85	160	140
G-6-P	160	250	250

Fig. 2. P^{32} in intermediates of glycogen breakdown. Liver slices were incubated with inorganic
radioactive phosphate with or without addition of
glycogenolytic agents. (from 4)

Well, after this, it was fairly simple to show that
the enzyme, phosphorylase, was activated by these hormones
in intact cells. Here we used dog liver slices, (Fig. 3)
taking the 0 time value arbitrarily as 100%. On incubation,
activity falls down to a low level. If epinephrine is
added at this stage, we can see what a tremendously rapid
recovery there is of phosphorylase activity. So this was
nice--and I think it helped me in my career quite a bit
because a lot of people thought this was great and I had
several heads of department offered me, and I finally
accepted one of them even though I wasn't too anxious--.
But then it turned out we couldn't go any further. We
were really stuck. We could activate the phosphorylase in
these intact cells but failed when we tried to break up
the cells and study this reaction so we could define it.
You know there are thousands and thousands of chemicals
and enzymes in these cells, and if you look from the outside, it is really very hard to decipher what is going on.
So we went for several years without much luck.

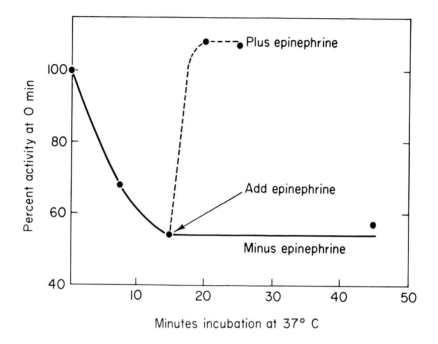

Fig. 3. Effect of epinephrine on the activity
of phosphorylase in dog liver slices (data from 5).

        This work with Dr. Wosilait (Fig. 4) I think provided
an important lead. We found that phosphorylase when it
changed in activity, did so because there was a change in
the phosphate group--and the figure shows one of our
earliest experiments that was published in this area. Here
we have enzyme activity falling with our inactivating
enzyme (abbreviated I.E.) which turned out to be a phos-
phatase--and a phosphate release accompanies this. This
was an early example of the effect of phosphorylation on
enzyme activity. So this pleased us quite a bit because
we thought it might be more complex than this -- and, you
know, sometimes you are lucky and sometimes you are not.

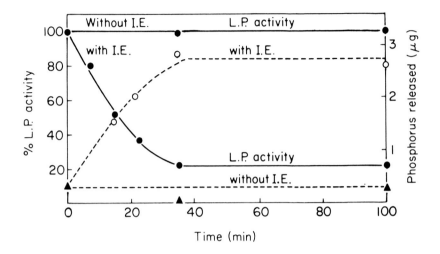

Fig. 4. Correlation of the loss of activity
with the release of phosphate from liver
phosphorylase (L.P.) brought about by an
"inactivating enzyme" (abbreviated I.E.)
subsequently shown to be a protein phosphatase.
(data from 6).

We had actually drawn up a list of guesses and an amide
group was our first guess and an amino group the second
and phosphate was third; but we had a reagent for phosphate
so we tried it first and it worked, and we were lucky there.

Now, working with Dr. Rall, we then went back to the
homogenate (see Fig. 5) and found that the hormones would
work; in fact, all we needed to do -- and I won't go into
the element of luck here -- was to fortify the homogenates
with magnesium ions and ATP which were necessary for this
conversion, and then, if we measured the enzyme formed, we
could see that the hormones did work. We could see micros-
copically that we had broken up the cells, and, furthermore,
we could add purified enzyme to this system and it too
would be activated. This showed conclusively we were not

9

dealing with intact cells. However, when we centrifuged
these preparations, even fairly gently, we lost our
hormone effect, and that puzzled us because ordinarily when
you make an homogenate you throw away all the debris, as it
is called. But when we threw away the debris--and it makes
it easier pipetting too--we lost our effect. When we added

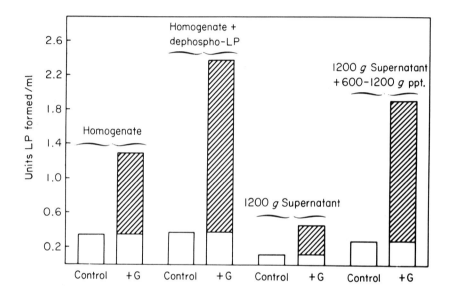

Fig. 5. Activation of liver phosphorylase (L.P.)
by glucagon (G) in a crude or fractionated homogenates of dog liver (from 7).

back some of the debris, the effect would return. So
what this really meant -- and will not go into detail --
is that the hormones do not act at the level of soluble
enzymes and the phosphorylase itself; instead, they react
with a particulate portion of the cell. And, as it finally
turned out, they made something that then did the work on
the enzyme.

And Fig. 6 shows the basic reaction that was studied here. It requires ATP, which is the basic fuel of the body, and, in the presence of the enzyme and magnesium ions, ATP is converted to a cyclic nucleotide and inorganic pyrophosphate. Now the body uses the cyclic AMP again after it is formed by opening the ring and tossing it right back in the metabolic cycle where it can go back to ATP. The enzymes involved are fairly specific and there have been several reports of hormone actions in the area of cyclic AMP degradation that still need confirmation. There are certainly a number of drugs that act in this area and many of the drug companies are concerned with this subject because you have the ability to increase cyclic AMP by an inhibitor and to decrease it if you can stimulate this reaction. Moreover it seems that this diesterase varies from organ to organ, and even within a tissue. Thus there is quite a lot of hope that chemicals can be found that will work on specific organs or cells. Actually, some of the compounds we are already using in medicine seem to act on the diesterase reaction. At least part of the action of

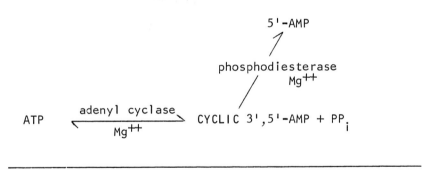

Fig. 6. Reactions involved in the formation and destruction of cyclic adenylate.

caffeine seems to be through this system and there are other drugs that are just now being studied like papaverine, which, you recall, relaxes smooth muscle, that also act through this system. So I predict that there will be much activity in this area in the future.

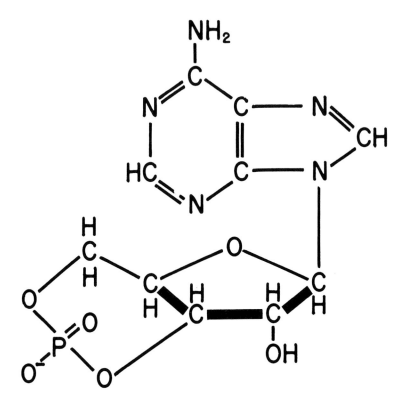

Fig. 7.   Structure of cyclic AMP (adenosine
3',5'-phosphate)

I won't discuss the structure of cyclic AMP in detail
(Fig. 7). It is simpler than it looks in some ways although it has four rings. It is noteworthy that the phosphate just doesn't dangle out in space, but curves back
on itself to make the 3' linkage. Now that linkage is
characteristic of ribonucleic acid, and in a way, this is
a little bit of ribonucleic acid. Some people call it sort
of a transistorized RNA.

Table I lists some of the effects of cyclic AMP, and,
although it is out of date, it is still long enough that I

can just point out some things I'll touch on later. I have
discussed phosphorylase activation already. This led to
the discovery of cyclic AMP. Now, the enzyme that brings
about the synthesis of glycogen is affected by a similar
mechanism except that its activity is turned off.

Table I

Effects of cyclic AMP (see 2,3)

Enzyme or process affected	Tissue	Change in activity or rate
Phosphorylase	Several	Increased
Glycogen Synthetase	"	Decreased
Protein kinase	"	Increased
Phosphofructokinase	Flatworm	"
Tyrosine transaminase induction	Liver	"
PEP carboxykinase induction	"	"
Serine dehydratase induction	"	"
β-Galactosidase induction	Bacteria	"

This makes a reciprocal system, you see. The enzyme that
breaks glycogen down is increased in activity whereas the
synthetic activity is decreased - a very nice system. The
activation of protein kinase by cyclic AMP, which has been
studied primarily by Ed Krebs' group, is in back of a lot
of these actions which we'll discuss. We don't know how
many cyclic AMP effects involve protein kinase because, in
some cases, we don't even know the enzymatic background,
the physiology, the biochemistry, or the reactions involved. Cyclic AMP is involved in enzyme inductions which
I may discuss a little later when we spend a couple of
minutes on bacteria.

Table II

Effects of cyclic AMP (see 2,3)

Enzyme or process affected	Tissue	Change in activity or rate
Overall protein synthesis	Several	Decreased
Gluconeogenesis	Liver	Increased
Ketogenesis	"	"
Lipolysis	Adipose	"
Steroidogenesis	Several	"
Water permeability	Epithelial	"
Ion Permeability	Several	"
Ca^{++} resorption	Bone	"

In Table II we have a series of different effects. Gluconeogenesis is stimulated, as shown by Park and Exton-- they are the real world experts in this. We and many others have worked on lipolysis, and I will show an experiment in this area and one on steroidogenesis where my group has worked with Dr. Liddle's group.

Water permeability is a very interesting thing and has to do with renal function, but we'll have to skip the others in the table.

Turning now to Table III, Al Robison did some nice work on the effect of cAMP on the contractility of heart muscle. cAMP also stimulates amylase release.

In Table IV, we have listed some of the hormones that are released by cyclic AMP. In at least two cases, the hormone is not only released by cyclic AMP but it then acts via the cyclic AMP mechanism. Now that is a little hard to understand in terms of feedback mechanisms. Ren-jye Ho is doing some nice work in the area of feedback mechanism which will, I think, develop very nicely in terms of control at the cellular level.

Table III

Effects of cyclic AMP (see 2,3)

Enzyme or process affected	Tissue	Change in activity or rate
Renin production	Kidney	Increased
Contractility	Cardiac muscle	"
Tension	Smooth muscle	Decreased
Membrane potential	Smooth muscle	Hyperpolarized
Amino acid uptake	Liver and uterus	Increased
" " "	Adipose	Decreased
Clearing factor lipase	"	"
Amylase release	Parotid	Increased
Insulin release	Pancreas	"

Table IV

Effects of cyclic AMP (see 2,3)

Enzyme or process affected	Tissue	Change in activity or rate
ACTH release	Anterior pituitary	Increased
TSH release	"	"
GH release	"	"
LH release	"	"
Thyroid hormone release	Thyroid	"
Calcitonin release	Parafollicular	"
Acetylcholine release	Nervous	"
Histamine release	Leucocytes	Decreased
HCl secretion	Gastric mucosa	Increased

15

I do want to mention two or three things in connection with Table V. The slime mold in its simplest form is a tiny amoebae which lives happily eating mostly bacteria. And it may be attracted to the bacteria by cyclic AMP, but that is debated. But more important I want to develop a few of the ideas about how the slime mold differentiates. A lot of biologists work in this area because this is a prime example of early differentiation. I'll try to summarize it briefly and indicate where cyclic AMP comes in at three steps. The little amoebae may be sitting in an environment that is favorable, in which case they will stay there and reproduce for a long time. When hard times come, however, some of the organisms start secreting a substance which for 30 or 40 years was called an acrasin. It has recently been shown by Bonner to be cyclic AMP. I may now simplify too much for the sake of covering ground, but, in the presence of cAMP, the little amoebae

Table V

Effects of cyclic AMP (see 2,3)

Enzyme or process affected	Tissue	Change in activity or rate
Fluid secretion	Insert salivary glands	Increased
Discharge frequency	Cerebellar Purkinje	Decreased
Melanocyte dispersion	Skin	Increased
Aggregation	Cellular slime mold	"
"	Platelets	Decreased
RNA synthesis	Bacterial	Increased
DNA synthesis	Thrombocytes	"
Cell growth	Tumor cells	Decreased

decides he has a head and a tail--and it takes him about 20 mins for this decision to be reached. Before this he has had no sense of head and tail. So this is the first step and I think it is a pretty big step. And now the second; in the presence of cAMP, the amoebae aggregate into a colony containing 100,000 to 200,000 organisms to form what is known as a slug. This is a kind of worm-like creature which contains all the amoebae joined together in an orderly fashion. The slug is able to walk out of the environment. And now the third, cAMP dependent step. The slug forms a big spike, at the top of which you get spore formation. These spores are very resistant to the environment and can be shed and will persist until times are better. Now I point this out because the area of differentiation has hardly been touched in biology since I was a student--and the above may be an important lead. Now, in the case of the platelet, Table V, as studied by Al Robison and others, aggregation is decreased by the formation of cAMP inside the cell. Finally, in Table V, we note the recent finding of Pastan from NIH, that certain types of tumor cells seem to revert to normal if you put cyclic AMP on the outside in the culture. And, in some cases, certain cell types like the lymphoma, seem even to die in the process. Now how far this will go and how much of this effect will be coupled with viruses I don't know but I would be surprised if next week there aren't a lot of scientists looking into this. Every year since I was a medical student there has been some dramatic, possible breakthrough in cancer, yet the problem is still with us. So I think you should look at this in the same light; it ought to be studied but where it will go we don't know yet.

Now I am going to show you an experiment with bacteria (Fig. 8). We had a student, Richard Makman, and he wanted to be a psychiatrist and yet he was interested in science, too. But more importantly, he had a wife who was a sophomore medical student, so he took a 2 year postdoctoral fellowship, and, of all things, worked on E. coli. In the beginning, we were interested primarily in growth. As shown in Fig. 8, E. coli were placed in a very simple medium containing ammonium salt and glucose. Growth was measured by the change in the turbidity (upper line). You see that they very quickly started to grow, but then this fantastic rise in cAMP occurred, which it turned out later, was related to the glucose concentration (broken line). We have never gone back and looked into the growth effects since then because the glucose effect was so overwhelming. Now, you see, the E. coli doesn't have as much

17

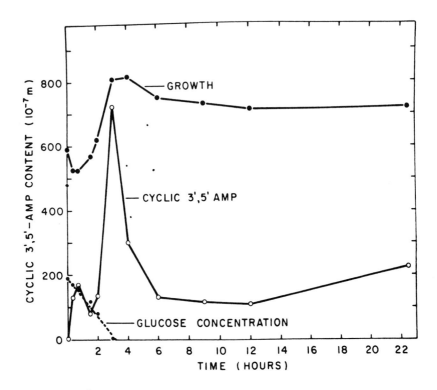

Fig. 8. Relationship of growth and glucose concentration to the cAMP content of E. coli (from 8)

space as mammals and it therefore, can't keep around all
sorts of enzymes. It has to make them when it needs them.
It turns out that cyclic 3',5'-AMP goes up when you remove
glucose. You can then get the induction of any of almost
20 enzymes if there is a trace of the appropriate inducing
substrate in the medium. This is related to so-called
glucose effect. So what does this mean? It means that
cyclic 3',5'-AMP is probably working at the level of both
transcription and translation. In other words, it is
active at the genetic level. This subject is being worked
on by a number of groups of workers because here is a
control that gets right down to the level of the operon.
Now in mammals, cAMP has an effect on induction too. I
don't think it has been studied in the embryo, but, in the

18

adult, you can bring about the induction of a limited
number of enzymes with cyclic AMP.

Table VI

Criteria for the Relation of

Cyclic AMP to Hormone Action

1. Demonstration of response to hormone in washed broken
cell preparations.

2 a. Intact cell concentrations of cyclic AMP should
change appropriately in response to hormone stimulation,
i.e. time etc.

2 b. Drugs which change phosphodiesterase activity should
potentiate or inhibit.

3. Exogenous cyclic AMP or a derivative (hopefully)
should mimic the effect of the hormone.

Table VI shows the criteria we developed to relate
cyclic AMP to hormone action. We have listed four points.
We like to work in washed broken cell preparations. They
are not always simpler, but sometimes they are. In fact,
we have never gotten some effects in broken cell preparations, but, if we can break the cells and wash out some
of the thousands of chemicals and enzymes that are around,
we feel we are closer to the primary effect and freer from
displacement effects. It is most important that the
changes within the cells should be appropriate in time, and
this is a very important criteria that can be used often
in the intact animals. The diesterase inhibitors or
activators can also occasionally be used in animals. Now,
finally, we would like to show that you can add exogenous
cAMP, or a derivative, and mimic the hormone effect.
cAMP and its derivatives do enter a few cells, but slowly,
since they are still pretty good size compounds with a
charge. I would judge by now that effects have been shown
on 10 - 20 different types of cells, but high concentrations of the nucleotides are usually needed. Some of the

derivatives are much more effective than cAMP itself.

When we studied the fat pad (Fig. 9)--and Bill Butcher working with Dr. Ren-jye Ho and Dr. Meng did a lot of this work--we could show that fatty acid was released in the medium with cyclic AMP, but, we wouldn't brag too much about this experiment. The most useful part about this experiment is to see how much more effective the derivative is. Dr. Posternak came from Switzerland and worked during his sabbatical year with us on these nucleotides. He is an inositol expert and this was his first experiment in nucleotides, but he made some very good compounds.

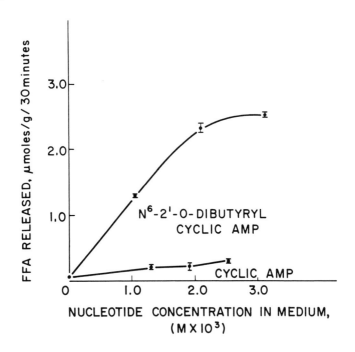

Fig. 9. The effect of N^6-2'-0-Dibutyryl cyclic AMP on FFA release by isolated fat cells (from 9).

Table VII

Dibutyryl 3',5'-cyclic AMP

as an inducer of enzyme secretion

Additions	Amylase secreted	
	15 min incubation	65 min incubation
	% of total	
None	1.5	4.7
Dibutyryl cyclic AMP, 1 mM	7.8	48.0
Dibutyryl cyclic AMP, 2 mM	10.0	51.0
Epinephrine, 0.01 mM	13.0	43.0

Now this is a little digression along the same lines (Table VII). Micky Schramm in Israel has been working with amylase release from the parotid gland and has found that epinephrine causes the release of amylase. This is a very interesting thing; and he has done quite a lot of histological work in this area. There are a lot of questions about membrane fusion and so forth, but he could not get cyclic AMP to do a thing in his system; it was completely negative. This was in the early days of cAMP research, and he wrote to ask if he could get a derivative and we did send him some dibutyryl cyclic AMP. He had very tiny test systems, so it did not take much, and you see it was very effective. This is an example of enzyme release in contrast to hormone release that we have discussed before. cAMP does lie back of this release of amylase, but how much it has to do with membrane fusion and the microtubular apparatus is not yet clear.

This is work on steroid release from the adrenal gland (Fig. 10) carried out with Bill Butcher in collaboration

21

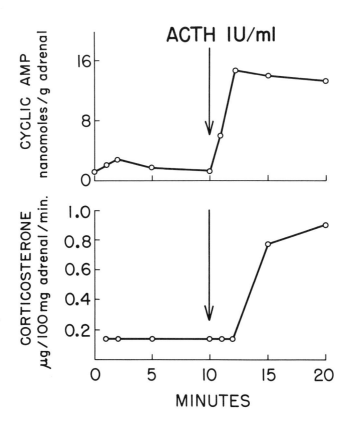

Fig. 10. Relationship in time between the ACTH stimulated rise in cAMP and the release of steroid by the adrenal gland (from 10).

with Dr. Grant Liddle's group at Vanderbilt. These studies were made in vitro, but similar results have been obtained in vivo. We are concentrating here on the time course and ACTH is given at the point shown by the arrow. You see an almost instantaneous rise in cAMP which preceeds any rise of steroids that we can detect. So, in the adrenal gland, all the criteria that we have listed have been met.

With the heart we had more trouble, but I think finally all the criteria have also been met, although we could not settle the last one ourselves. The work shown in Fig. 11 is by Al Robison and other colleagues. We have perfused working rat hearts. A little epinephrine was squirted into the perfusion tube just proximal to the heart in suboptimal amounts, and then, after a certain time in seconds, the tissue was rapidly frozen and cyclic AMP was measured. All criteria were met that the epinephrine effect was mediated by cAMP although we ourselves could not find an effect of exogenous cyclic AMP or derivatives. Since that time others have set up the heart preparations in different ways and in different species and have shown that the derivatives of cyclic AMP can increase the contraction force. So I think that it is now pretty well accepted that when your heart beats stronger due to epinephrine, and probably also faster, that cyclic AMP is back of those phenomena.

Now I'm going to show you an area where there is a lot of speculation in the field of medicine and not too much has been done about it. Let me point out a few things briefly. Studies on the action of insulin have gone forth in our lab in collaboration with Dr. Park and his group and it is clear that insulin can lower the level of cyclic AMP in fat and liver. I think this work is well established, but the effect does not necessarily explain all the actions of insulin. Now the release of insulin is also related to cyclic AMP. Most of you know that a rise in blood sugar is really what calls out the release of insulin, but cyclic AMP is operating at the same time in this release. We come here to another point which has been worked out largely by Cerasi and Luft in Sweden because they have a registry of identical twins. You know that is a nice thing to have. It would be nice if this country had one because it would find many applications to problems of genetics. Identical twins have the same genes--and you know the chance of two people at random having the same genes is one in several trillion. So these twins provide a chance to work on a number of things from psychological to physical problems. Anyway, Cerasi and Luft found 10 identical twins in which only one individual in each pair had come down with diabetes. Nevertheless, it was certain that the genetic defect was present in the other twin.

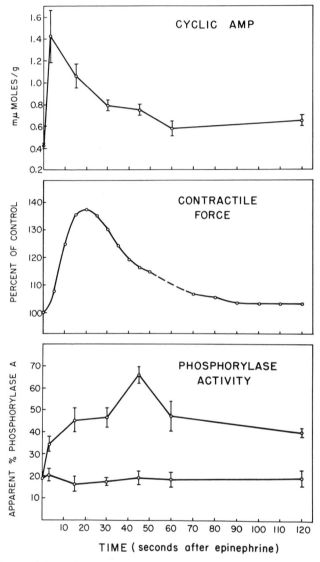

Fig. 11. Time relationships of the epinephrine induced changes in cAMP, contractile force and phosphorylase activity in the perfused rat heart. (The lower curve in the bottom panel is the activity without catecholamines (from 11)

Now, one out of five of you here in this audience will also
have this same genetic defect, but very few will come down
with diabetes because there are other unknown precipitating
causes that are very important. In the Cerasi and Luft
series, the diabetes free twin did not even have an
abnormal glucose tolerance test after a steroid challenge.
Nevertheless, it was found in these individuals that there
was a lag in insulin secretion and that this lag could be
corrected by an inhibitor of the cAMP phosphodiesterase.
They also went on with other studies to show that there is
a tonic depressing effect of the catecholamines on insulin
release. These studies in the area of insulin release are
very tough. You have to work with small amounts of tissue
and even isolated islets are not satisfactory. We need to
have a preparation of just beta cells alone. This is a
place where methods need to be refined. In general, we
spend about a third of our time working on methods--all
the way down the line--from enzymes to cultures--we need
better and better methods.

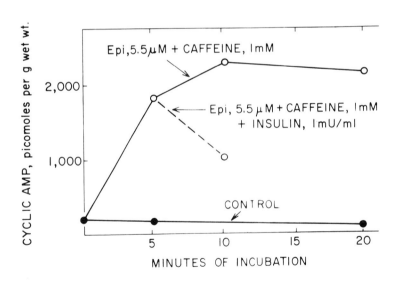

Fig. 12. Effect of insulin on the level of cAMP
in the rat epididymal fat pad (from 12)

Fig. 12 is just to show the lowering effect of insulin on cyclic AMP in fat. We've incubated rat epididymal fat pads with both epinephrine and caffeine to get tremendously high levels of cAMP. When insulin was added, down went the level of cyclic AMP.

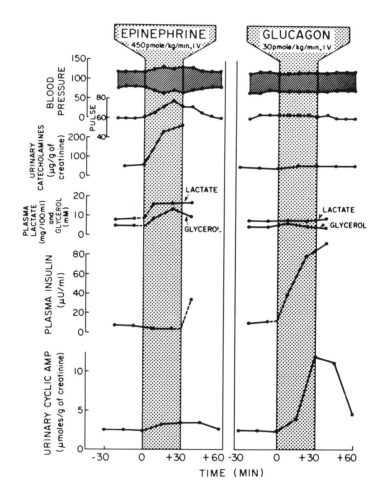

Fig. 13. Contrasting effects of epinephrine and glucagon on insulin secretion and other parameters in normal man (from 13)

26

The work shown in Fig. 13 was carried out by Arthur Broadus and several of Dr. Liddle's colleagues using normal men as subjects. The study was not set up to study the release of insulin and most release work in fact has been done by others. I could name five groups, but I will not go into it now. But I did have this figure handy, and, while Broadus' group was after another point, the study illustrates the point I want to make. Epinephrine was injected where shown and the blood glucose of course shoots up, and it does the same with glucagon. The phenomenon is so well recognized that it is not even shown on the slide. You see that the plasma insulin actually falls after epinephrine, even in the face of a rising blood sugar. However, in the case of glucagon, there is an effect to raise the plasma glucose, but also a positive effect on cyclic AMP itself - so up shoots the insulin secretion. Now these are regulations operating in normal men. How far studies with human subjects can be pursued and how difficult they will be, I don't know, but this may indicate to you one of the areas where work is being done. A lot more has been done in man in the parahormone area, but I will not review that here.

I want to mention to you that there is another cyclic nucleotide in nature, cyclic guanylate. This was found by Price in rat urine. Price was a relatively young man who died an early death. We had also been interested in this compound and were planning to work on it. We had several theories about how it might be effective in the action of insulin and several other things that have never panned out, but that haven't been completely disproved yet, either. In table VIII, an experiment is shown that Dr. Hardman set up with Jim Davis. They simply took the pituitary out of rats--in fact, they didn't even do that, since they could buy them with the pituitary already out. They then compared the operated with the normal rats and, as you can see, the cyclic GMP in the urine of the normal rat is high. There is also a lot of cyclic AMP that floats out in the urine, and there is some in the blood stream. There is also cyclic GMP in the blood. Now, the cyclic GMP in urine falls quite a bit when you take out the pituitary while cyclic AMP falls only a little bit. But, with restoration of urine volume back to normal with cortisol, cAMP comes back to normal, but cGMP does not.

Thus, the two nucleotides appear to be controlled separately.

Table VIII

Excretion of Cyclic GMP and Cyclic AMP in Urine

from Normal and Hypophysectomized Rats

	Cyclic GMP nmoles/100 gm rat	Cyclic AMP nmoles/100 gm rat	Urine Volume
Normal	28 ± 1.3 (30)	59 ± 2.6 (26)	38 ± 1.9 (30)
Hyphex	13 ± 1.0 (22)	49 ± 3.8 (19)	13 ± 1.2 (22)
Hyphex + Hydrocortisone	12 ± 0.9 (23)	60 ± 4.6 (20)	32 ± 2.0 (23)

(data taken from 14)

As seen in Fig. 14, the normal rat secretes a certain amount of cyclic GMP, but with the pituitary out, the level falls. With hydrocortisone, it does not return to normal, but, if you inject 6 pituitary hormones together, it comes back to normal and perhaps even overshoots. Now, this has not been broken down well. Each time we take out one of the 6 hormones, the level drops. We still don't really know what the main factors are in this phenomenon except that the thyroid and the adrenals play a role. However, there is something else besides. This is a big area that Drs. Hardman and Schultz are working on. cGMP is present in all our cells too. I believe about 40 to 50 percent of our effort is now devoted to working on cyclic guanylate, which we hope someday will also be an interesting compound in both health and disease.

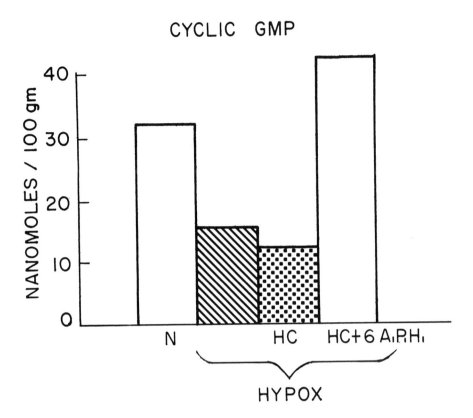

Fig. 14. Effect of certain hormones on the excretion
of cyclic guanylate in the urine of rats. N indicates
normal; hypox indicates hypophysectomized; HC
indicates hydrocortisone; and 6 A.P.H. indicates six
anterior pituitary hormones (G.H., ACTH, TSH, FSH,
LH and lactogenic hormone)
(data taken from 14).

1. Sutherland, E.W., Science 177 (1972) 401.
2. Robison, G.W., Butcher, R.W., and Sutherland, E.W. Cyclic AMP, Academic Press, 1971.
3. Sutherland, E.W., and Robison, G.A., Pharmacological Rev. 18 (1966) 145.
4. Sutherland, E.W. and Cori, C.F., J. Biol. Chem. 188 (1951) 531.
5. Rall, T.W., Sutherland, E.W. and Wosilait, W.D., J. Biol. Chem. 218 (1956) 483.
6. Wosilait, W.D., and Sutherland, E.W., J. Biol. Chem. 218 (1956) 469.
7. .Rall, T.W. and Sutherland, E.W. and Berthet, J., J. Biol. Chem. 224 (1957) 463.
8. Makman, R.S. and Sutherland, E.W., J. Biol. Chem. 240 (1965) 1309.
9. Meng, H.C. and Ho, R.J., Progr. Biochem. Pharmacol. 3 (1967) 207.
10. Grahame-Smith, D.G., Butcher, R.W., Ney, R.L. and Sutherland, E.W., J. Biol. Chem. 242 (1967) 5535.
11. Robison, G.A., Butcher, R.W., Oye, I., Morgan, H.E. and Sutherland, E.W., Mol. Pharmacol. 1 (1965) 168.
12. Butcher, R.W., Sneyd, J.G.T., Park, C.R. and Sutherland, E.W., J. Biol. Chem. 241 (1966) 1652.
13. Broadus, A.E., Kaminsky, N.I., Northcutt, R.C., Hardman, J.G., Sutherland, E.W. and Liddle, G.W., J. Clin. Invest. 49 (1970) 2237.
14. Hardman, J.G., Davis, J.W. and Sutherland, E.W., J. Biol. Chem., 244 (1969) 6354.

THE REGULATION OF MUSCLE METABOLISM AND FUNCTION BY PROTEIN PHOSPHORYLATION

E. G. KREBS, J. T. STULL, P. J. ENGLAND
T. S. HUANG, C. O. BROSTROM and
J. R. VANDENHEEDE
Department of Biological Chemistry
University of California School of Medicine
Davis, California 95616

Abstract: Rabbit skeletal muscle phosphorylase kinase is a Ca^{2+}-requiring protein kinase, which, in addition to catalyzing the phosphorylation of phosphorylase b, also catalyzes the phosphorylation of troponin subunits. The dephosphorylation of these components is catalyzed by phosphorylase phosphatase. The thesis that phosphorylase kinase should be looked upon as a general Ca^{2+}-requiring protein kinase is examined.

INTRODUCTION

Skeletal muscle phosphorylase kinase catalyzes the conversion of phosphorylase b to phosphorylase a as shown in Equation 1:

$$2 \text{ Phosphorylase b} + 4 \text{ ATP} \xrightarrow{\text{Phosphorylase Kinase}} \text{Phosphorylase a} + 4 \text{ ADP} \qquad (1)$$

The reversal of this process is catalyzed by phosphorylase phosphatase:

$$\text{Phosphorylase a} + 4 \text{ H}_2\text{O} \xrightarrow{\text{Phosphorylase Phosphatase}} 2 \text{ Phosphorylase b} + 4 \text{ Pi} \qquad (2)$$

These reactions and their counterpart in liver represent the first reported examples of the process of protein phosphorylation-dephosphorylation as a mechanism for metabolic control (1,2). Other examples were reported in due course (3,4) and the very fact that an entire symposium can now be built around this subject attests to the importance and generality of this type of regulatory system. Interest in protein

31

phosphorylation reactions was greatly enhanced by the finding that cyclic AMP acts by stimulating a protein kinase (5).

THE REQUIREMENT OF PHOSPHORYLASE KINASE FOR CALCIUM

Phosphorylase kinase is a Ca^{2+}-requiring enzyme. This fact was first appreciated when it was found that its activity was strongly inhibited by the chelating agents, EDTA and EGTA, and that this inhibition was reversed by addition of calcium salts (6). This is illustrated in Table I for an experi-

TABLE 1

Effectiveness of various metal ions in reversing the inhibition of rabbit skeletal muscle phosphorylase kinase by EDTA*

Sufficient EDTA, 1 mM, was added to cause 60 percent inhibition of the phosphorylase b to phosphorylase a reaction. The concentration of Mg^{2+} was 10 mM and the ATP concentration was 3 mM. Salts of the different metals were added at a final concentration of 0.5 mM. pH = 8.2

Metal Ion	Inhibition of kinase activity (%)
None	60
Ca^{2+}	4
Sr^{2+}	19
Mn^{2+}	37
Ba^{2+}	44
Co^{2+}	48
Fe^{2+}	52
Zn^{2+}	59
Cu^{2+}	82

*Table reproduced in part from Meyer et al. (6) by permission of the American Society of Biological Chemists, Inc.

ment utilizing EDTA as the inhibitory agent. Ozawa et al. (7) put the Ca^{2+} requirement on a more quantitative basis and showed that the sensitivity of the kinase to Ca^{2+} was sufficiently high that its activity could conceivably be controlled in vivo by this ion. These workers estimated that half maximal stimulation of the kinase would occur in the micromolar range of free Ca^{2+}. Brostrom et al. (8), utilizing calcium-free reagents, demonstrated a requirement of the enzyme for this metal directly. The binding of ^{45}Ca to the enzyme was also studied. It was demonstrated that the activated form of phosphorylase kinase (see below) appears to be slightly more sensitive to Ca^{2+} than the nonactivated form.

THE PHYSIOLOGICAL ROLE FOR CALCIUM

The requirement of muscle phosphorylase kinase for calcium is believed to relate to its control with respect to the coupling of glycogenolysis to muscle contraction (6-12). A plausible concept is that the release of Ca^{2+} from the sarcoplasmic reticulum not only causes contraction of the myofibril but also triggers glycogenolysis through the effect of the metal on phosphorylase kinase. The scheme is presented diagramatically in Figure 1. Inasmuch as muscle contraction

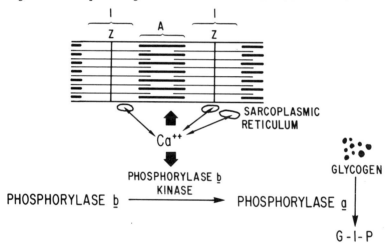

Fig. 1. The coupling of muscle contraction and glycogenolysis. Reproduced from Brostrom et al (8) by permission of the American Society of Biological Chemists, Inc.

requires ATP and glycogenolysis (glycolysis) yields ATP, the regulation of these two processes by a common messenger makes good sense. As noted earlier, the high affinity of phosphorylase kinase for Ca^{2+} (7) is in keeping with the ability of the metal to act in this capacity. A direct demonstration of the validity of the scheme shown in Figure 1 was carried out by Brostrom et al. (8) who showed that an isolated sarcoplasmic reticulum preparation is capable of inhibiting phosphorylase kinase in vitro.

FORMS OF PHOSPHORYLASE KINASE AND THE SIGNIFICANCE OF THEIR INTERCONVERSION

Rabbit skeletal muscle phosphorylase kinase is known to exist in two forms commonly referred to as nonactivated and activated forms of the enzyme (13-15). Nonactivated phosphorylase kinase is a dephospho form and activated phosphorylase kinase is a phosphorylated form. Both forms require Ca^{2+}. Their pH optima are presented in Figure 2. The conversion of

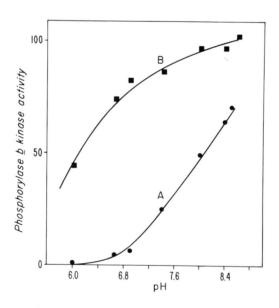

Fig. 2. The pH optima for nonactivated (Curve A) and activated (Curve B) rabbit skeletal muscle phosphorylase kinase. Maximal activity is taken as 100. Reproduced from Krebs et al. (15) by permission of the Williams and Wilkins Co.

nonactivated phosphorylase kinase to its activated form is catalyzed by the cyclic AMP-dependent protein kinase (5). This latter kinase was first discovered and appreciated as having a broad specificity in protein phosphorylation reactions as a result of work on the catalysis of the phosphorylase kinase activation reaction. The enzyme had been recognized as a catalytic entity, however, by Friedman and Larner (14) who utilized it as glycogen synthetase kinase.

The physiological role of phosphorylase kinase activation in muscle is not entirely clear. Although it has been generally accepted (15-18) that this process constitutes an integral step in the chain of events which occur when epinephrine stimulates glycogenolysis, recent experiments have shown that small doses of isoproterenol are capable of causing phosphorylase a formation in muscle without a concomitant activation of phosphorylase kinase (19). The formation of phosphorylase a which accompanies muscle contraction induced by electrical stimulation is apparently not due to phosphorylase kinase activation (17, 19), nor is there an increase in the levels of cyclic AMP in muscle stimulated electrically (16).

It should be noted in passing, that nonactivated phosphorylase kinase can be converted to its activated form by limited proteolysis as well as by enzymatic phosphorylation (6,20). This type of activation is irreversible and is not believed to be of physiological significance with reference to the regulation of glycogenolysis. Skeletal muscle and other tissue (21) contain a Ca^{2+}-requiring protease, originally referred to as the kinase activating factor (KAF). At the time that KAF was first described (6), it was also noted that the reversible stimulation of KAF-free phosphorylase kinase by Ca^{2+} also occurs.

THE ENZYMATIC PHOSPHORYLATION OF TROPONIN

Troponin, a protein complex found in association with the actomyosin of muscle, is essential for the sensitivity of actomyosin ATPase to Ca^{2+} (see reference 22 for appropriate citations). During the past year several reports have appeared which showed that troponin can be phosphorylated in vitro utilizing different protein kinases to catalyze the reaction(s). Bailey and Villar-Palasi (23), who were the first to describe this phenomenon, demonstrated that the cyclic AMP dependent protein kinase was effective in this reaction. In their experiments the inhibitory subunit of troponin (TN-I) was phosphorylated whereas neither the calcium building subunit (TN-C) nor the tropomyosin binding subunit (TN-T) were phosphorylated. In this laboratory (24) it was found that TN-I was phosphorylated by phosphorylase kinase as well as by cyclic AMP-dependent protein kinase. Moreover, the phosphorylated form of TN-I could also be dephosphorylated in a rapid reaction catalyzed by phosphorylase phosphatase, an enzyme which heretofore had been thought to be entirely specific for phosphorylase a (25). The possible significance of the interconversions of phospho and dephospho forms of TN-I by phosphorylase kinase and phosphorylase phosphatase is further enhanced by consideration of the results of kinetic analyses (Table 2).

TABLE 2

Comparison of phosphorylation and dephosphorylation
of TN-I and phosphorylase

Phosphorylase kinase measurements were made at
pH 7 with nonactivated phosphorylase kinase as previously
described (24). Phosphorylase phosphatase values are from
England et al. (25).

| | Phosphorylase Kinase | |
	Km	Relative Vmax
Phosphorylase b	370 μM	11
TN-I	5.4 μM	1

| | Phosphorylase Phosphatase | |
	Km	Relative Vmax
Phosphorylase a	7 μM	0.25
Phospho TN-I	21 μM	1

The low Km values and appreciable Vmax rates of phos-

phorylase kinase and phosphorylase phosphatase when TN-I
is the substrate suggests that these reactions may be of importance in vivo. Pratje and Heilmeyer (26) found that the
TN-T subunit of troponin could serve as a substrate for the
cyclic AMP-dependent protein kinase, and recently it has
been determined in this laboratory that TN-T is also a substrate for phosphorylase kinase. In addition, TN-T phosphory
lated by phosphorylase kinase is dephosphorylated by phosphorylase phosphatase. Thus, it appears that Ca^{2+} stimulatior
of phosphorylase kinase activity in muscle may result not only
in phosphorylase a formation, but also in phosphorylation of
troponin (Figure 3).

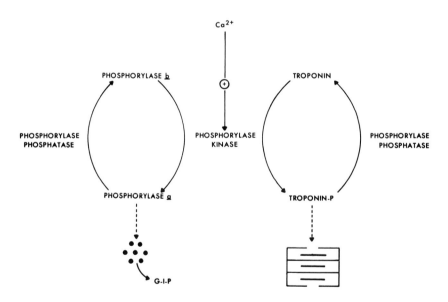

Fig. 3. Diagram of the relationship between phosphorylase and troponin phosphorylation.

The physiological significance of troponin phosphorylation has not been determined, but it is of interest that freshly
isolated troponin contains endogenous bound phosphorous (25)
indicating that the phosphorylation of the protein apparently
occurs in vivo. If it is postulated that the inhibitory effect of
troponin on actomyosin is affected by its state of phosphorylation, then a number of interesting interrelationships emerge.
One of these is that calcium may act at two distinct regulatory
sites in the control of muscle contraction. One of these sites
would be the stimulation of phosphorylase kinase and its subsequent catalysis of troponin phosphorylation, and the other
would be in the direct effect of this metal on the interaction of
troponin-tropomyosin with actin (22). A second relationship
would concern the role of cyclic AMP as a regulatory agent
though its effect on the stimulation of the cyclic AMP-dependent
protein kinase. The question could even be raised as to
whether or not the Ca^{2+}-dependent phosphorylation of troponin
by phosphorylase kinase would not in turn be affected by the
cyclic AMP-dependent phosphorylation of phosphorylase kinase
by the protein kinase. This latter complication would not seem
to enter into the picture inasmuch as nonactivated and activated
phosphorylase kinase manifest the same activity toward TN-I
(Figure 4).

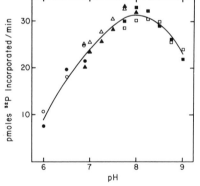

Fig. 4. The effect of pH on the phosphorylation of troponin B (fraction which contains TN-I and TN-T) by phosphorylase kinase. The closed symbols indicate activated phosphorylase kinase and the open symbols nonactivated kinase.
Circles, triangles and squares indicate different buffers. Reproduced from Stull et al. (24) by permission of the American
Society of Biological Chemists, Inc.

PHOSPHORYLASE KINASE AS A CALCIUM-DEPENDENT PROTEIN KINASE OF GENERAL SIGNIFICANCE

In addition to catalyzing the phosphorylation of phosphorylase and troponin, this enzyme also catalyzes the Ca^{2+}-dependent phosphorylation of casein albeit at a slow rate (27). Phosphorylase kinase also catalyzes its own phosphorylation in what has been referred to as an autoactivation process (27). Thus, disregarding the question of the physiological significance of these reactions, it could be said that four substrates for the enzyme have been identified. On this basis a general scheme for the phosphorylation of regulatory proteins by this Ca^{2+}-dependent enzyme can be postulated. Such a scheme is presented in Fig. 5.

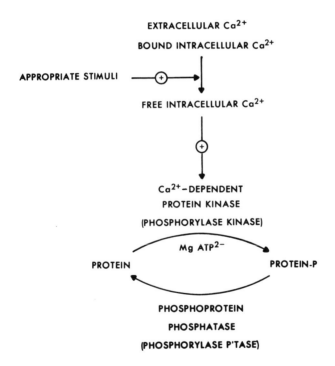

Fig. 5. A general scheme for the phosphorylation and dephosphorylation of protein by a Ca^{2+}-dependent mechanism with phosphorylase kinase and phosphorylase phosphatase.

Although there would appear to be some overlapping specificity of phosphorylase kinase and the cyclic AMP-dependent protein kinase with reference to the phosphorylation of troponin subunits, a subject area clearly in need of further work, the two enzymes have distinctly different specificities with respect to other substrates. The cyclic AMP-dependent enzyme does not catalyze the phosphorylation of phosphorylase. Histone is not phosphorylated by phosphorylase kinase.* The rates of phosphorylation of casein by the cyclic AMP-dependent protein kinase and by phosphorylase kinase are clearly different based on turnover numbers (not illustrated). Thus, these two enzymes would provide for a variety of protein phosphorylation reactions in a single cell type based on the nature of the messenger, i.e. Ca^{2+} or cyclic AMP, causing activation of the phosphorylation system.

A preliminary tissue survey for Ca^{2+}-dependent protein phosphorylation reactions has been carried out in rabbits utilizing skeletal muscle phosphorylase b as a substrate. In the results obtained thus far, only heart, brain, and skeletal muscle have shown the presence of such an enzyme as tested in crude extracts. Liver and kidney did not appear to contain a Ca^{2+}-dependent enzyme active toward phosphorylase b, although phosphorylase kinase activity was, of course, demonstrable in these tissues. In this connection it is of interest that Khoo et al. (28) found that adipocyte phosphorylase kinase was Ca^{2+}-independent with respect to its phosphorylation of adipocyte phosphorylase but was partially Ca^{2+}-dependent with respect to the phosphorylation of skeletal muscle phosphorylase. It is apparent that the requirements for the activator may be related to the nature of the protein substrate utilized.

*Phosphorylase kinase preparations are usually contaminated with traces of the cyclic AMP-dependent protein kinase which probably accounts for a very slow reaction that is sometimes seen with histone as a substrate.

REFERENCES

(1)   Krebs, E. G. and E. H. Fischer, Biochim. Biophys.
        Acta, 20 (1956) 150.

(2)   Rall, T. W., E. W. Sutherland, and W. D. Wosilait,
        J. Biol. Chem., 218 (1956) 483.

(3)   Friedman, D. L. and J. Larner, Biochemistry 2 (1963)
        669.

(4)   Langan, T. A., in: Regulatory Mechanisms for Protein
        Synthesis in Mammalian Cells, eds. A. San Pietro,
        M. R. Lamborg, and F. T. Kenney (Academic Press,
        New York, 1968) p. 101.

(5)   Walsh, D. A., J. P. Perkins, and E. G. Krebs, J. Biol.
        Chem. 243 (1968) 3763.

(6)   Meyer, W. L., E. H. Fischer, and E. G. Krebs,
        Biochemistry, 3 (1964) 1033.

(7)   Ozawa, E., K. Hosoi, and S. Ebashi, J. Biochem.
        (Tokyo), 61 (1967) 531.

(8)   Brostrom, C. O., F. L. Hunkeler, and E. G. Krebs,
        J. Biol. Chem., 246 (1971) 1961.

(9)   Drummond, G. I., J. P. Harwood, and C. A. Powell,
        J. Biol. Chem., 244 (1969) 4235.

(10)   Heilmeyer, L. M. G., F. Meyer, R. H. Haschke, and
        E. H. Fischer, J. Biol. Chem., 245 (1970) 6649.

(11)   Mayer, S. E., D. H. Namm, and J. P. Hickenbottom,
        Advan. Enzyme Regul., 8 (1970) 205.

(12)   Villar-Palasi, C., and S. H. Wei, Proc. Nat. Acad. Sci.
        U.S., 67 (1970) 345.

(13)   Krebs, E. G., D. J. Graves, and E. H. Fischer, J. Biol.
        Chem., 234 (1959) 2867.

(14)   Krebs, E. G., D. S. Love, G. E. Bratvold, K. A. Trayser, W. L. Meyer, and E. H. Fischer, Biochemistry, 3 (1964) 1022.

(15)   Krebs, E. G., R. J. DeLange, R. G. Kemp, and W. D. Riley, Pharmacol. Rev., 18 (1966) 163.

(16)   Posner, J. B., R. Stern, and E. G. Krebs, J. Biol. Chem., 240 (1965) 982.

(17)   Drummond, G. I., J. P. Harwood, and C. A. Powell, J. Biol. Chem., 244 (1969) 4235.

(18)   Lyon, J. B., Jr. and S. E. Mayer, Biochem. Biophys. Res. Commun., 34 (1969) 459.

(19)   Stull, J. T. and S. E. Mayer, J. Biol. Chem., 246 (1971) 5716.

(20)   Huston, R. B. and E. G. Krebs, Biochemistry, 7 (1968) 2116.

(21)   Drummond, G. I. and L. Duncan, J. Biol. Chem., 241 (1966) 3097.

(22)   Katz, A. M., Physiol. Rev., 50 (1970) 63.

(23)   Bailey, C. and C. Villar-Palasi, Fed. Proc., 30 (1971) 1147.

(24)   Stull, J. T., C. O. Brostrom, and E. G. Krebs, J. Biol. Chem. 247 (1972) 5272.

(25)   England, P. J., J. T. Stull, and E. G. Krebs, J. Biol. Chem., 247 (1972) 5275.

(26)   Pratje, E. and L. M. G. Heilmeyer, Jr., FEBS Letters, 27 (1972) 89.

(27)   DeLange, R. J., R. G. Kemp, W. D. Riley, R. A. Cooper, and E. G. Krebs, J. Biol. Chem. 243 (1968) 2200.

(28)   Khoo, J. C., D. Steinberg, B. Thompson, and S. E. Mayer, J. Biol. Chem., in press.

The authors wish to acknowledge the support of the Muscular Dystrophy Associations of America, the National Institutes of Arthritis and Metabolic Diseases, NIH, U.S. Public Health Service (AM 12842) and the Damon Runyon Foundation in this work.

DISCUSSION

F. HUIJING: I would like some clarification as to the role of these different metals. In my experience the magnesium ATP chelate is always the substrate, and in addition, free magnesium ions are required for the phosphorylase kinase in rather high concentrations (L.B. Clerch and F. Huijing, Biochim. Biophys. Acta 268 (1972) 654). When you say that these other metals are also giving activity, do you mean that they replace the micromolar requirement of calcium or requirement of free magnesium or can they replace magnesium when it is chelated to ATP so could we have a strontium ATP chelate that is actually the substrate?

E.G. KREBS: There are three types of metal effects on phosphorylase kinase. The first is the formation of the magnesium ATP complex, which is probably the real substrate for the phosphorylation. The second effect, which you mentioned is that the magnesium ion at concentrations above that required to form the magnesium ATP complex also has an effect on enzyme activity. Neither of these effects, however, is related to calcium effect. This shown by the fact that the chelating agent, EGTA, which does not bind magnesium will completely block the activity of phosphorylase kinase even in the presence of a very high magnesium concentration.

B.L. HORECKER: I was interested in your suggestion that the C-subunit of phosphorylase kinase might be the catalytic subunit, while the others are the regulatory subunits, and that you have not been able to separate the subunits even at high dilution. Did you try this in the presence of calcium, which might be expected to loosen the combina-

43

tion of the catalytic and regulatory subunits?

E.G. KREBS: We have not been able to dissociate the enzyme reversibly under any conditions. Even the presence of calcium appears to have no effect on the dissociation.

O.M. ROSEN: Is phosphorylase kinase able to catalyse the phosphorylation of histones and protamine as well?

E.G. KREBS: Not at a significant rate.

D. STEINBERG: I was interested that the pH 6.8 to 8.2 ratio in some experiments doesn't seem to go up even though phosphorylase kinase seems to be undergoing phosphorylation and presumably activation. I was especially interested because in adipose tissue after exposure to epinephrine, we do not see an activation of phosphorylase kinase, i.e. an increase in the pH 6.8:8.2 ratio under conditions that lead to an activation of phosphorylation in the same experiments. We considered the possibility that the phosphorylase activation was an artifact, occurring during homogenisation. Could you comment on the difference between these experiments where the ratio does not change, and the earlier experiments where you showed that it does?

E.G. KREBS: In your own experiments with adipose tissue phosphorylase kinase, were you using muscle phosphorylase b as a substrate or adipose tissue phosphorylase?

D. STEINBERG: We used both in some experiments we added muscle phosphorylated b, in other experiments it was the endogenous phosphorylase that was being used as a substrate and just following the increase in the endogenous phosphorylase activity.

E.G. KREBS: The experiments with troponin suggests that the nature of the substrate is very important in determining this pH 6.5:8.2 ratio. With all the other substrates, phosphorylase b, phosphorylase kinase, and casein, we do see an effect of activation on the ratio.

R. PIRAS: Since there are indications that in vivo calcium may play a role in the regulation of muscle glycogen synthetase (R. Staneloni and R. Piras, Biochem Biophys. Res.

Comm. 36 (1969) 1032 and C. Vilches, M.M. Piras and R. Piras Molec. Pharmacol. 8 (1972) 780), I wonder whether you find an effect of calcium in the phosphorylation of glycogen synthetase by protein kinase?

K.G. KREBS: I am not aware that anyone has found any effect. I know that the chelating agent EGTA has been used in a study of the glycogen synthetase phosphorylase and my recollection is that it has no effect. This would seem to rule out calcium as any sort of an effector on that reaction.

F. HUIJING: I can confirm that cyclic AMP dependent protein kinase just like the phosphorylase kinase requires free magnesium ions in addition to the magnesium ATP^{2-} but no calcium.

S.A. ASSAF: You mentioned that phosphorylase kinase has been studied in liver but you did not elaborate. We are studying phosphorylase kinase in other tissues than muscle and we do not see an activation by calcium. Would you comment on that please?

E.G. KREBS: The phosphorylase kinase of liver does not appear to be a calcium requiring enzyme. The enzymes from heart muscle and skeletal muscle are calcium requiring enzymes. At this point I would really rather not comment on the kinase from other tissues, but they seem to fall into two patterns, some resemble liver, some resemble muscle.

S.H. APPEL: You noted that the C-component is the one that binds calcium and yet in your studies it is the only one that is not phosphorylated with ^{32}P. Does it have cold-phosphate on it that might be binding the calcium?

E.G. KREBS: Our studies of endogenous phosphate in phosphorylase kinase have not shown any in the C-component.

HORMONALLY REGULATED ENZYMES IN ADIPOSE TISSUE LINKED TO CYCLIC AMP-DEPENDENT PROTEIN KINASE

D. STEINBERG
Department of Medicine
School of Medicine, University of California, San Diego

Abstract: Evidence that hormone-sensitive lipase in rat adipose tissue is activated and phosphorylated by cAMP-dependent protein kinase is reviewed. The enzyme in human adipose tissue was shown to have similar properties and to be similarly activated. The effects of epinephrine on three systems in rat adipocytes—lipase, phosphorylase and glycogen synthase—were found to parallel cAMP production. Insulin counteracted the effects of epinephrine on these enzymes but did not prevent the epinephrine-stimulated increase in cAMP. The general question of regulation of cAMP-linked systems at "post-cAMP" levels is discussed with particular reference to the site of action of insulin.

INTRODUCTION

The rate-limiting step in mobilization of free fatty acids (FFA) from adipose tissue is the hydrolysis of the first ester bond catalyzed by hormone-sensitive lipase (HSL), a true triglyceride lipase (1-3). Adipose tissue also contains high levels of diglyceride and monoglyceride lipolytic activity but these activities are not stimulated even under conditions of hormone exposure that increase triglyceride lipolytic activity severalfold (1,2,4). The several fast-acting hormones that rapidly increase HSL activity (including epinephrine, glucagon and ACTH) simultaneously

increase phosphorylase activity in adipose tissue (5-9) and decrease glycogen synthase activity (8,10). As discussed further
below, phosphorylation has been directly demonstrated to accompany HSL activation (11,12). Changes in phosphorylase kinase
activity have not been directly demonstrated but the responsiveness of phosphorylase to hormonal stimulation and the parallelism
between cAMP formation and phosphorylase activation (10), discussed further below, implies the presence of a system analogous
to that in skeletal muscle (13). Glycogen synthase again has
not been purified from adipose tissue and the case for covalent
modification here rests on analogy with the system in muscle and
in liver (14,15). Thus, adipose tissue contains at least three enzyme systems believed to be regulated by covalent modification,
specifically via phosphorylation catalyzed by cAMP-dependent
protein kinase (PK) (11,12,16-19). If it were established that
all three enzyme systems in adipose tissue are in fact controlled
by cAMP-dependent protein kinase, the adipocyte would represent a useful model system in which to explore the interesting
question of regulatory sites beyond cAMP. In view of the mani-
fold systems regulated by cAMP one must ask whether all such
cAMP-regulated systems are automatically and directly triggered
once the hormone has activated adenyl cyclase. The list of
cAMP-mediated effects is long and varied, including effects on
cell growth, transcription, translation and many enzymes — including at least three in adipocytes. Regulation by alterations in
phosphodiesterase activity will be recognized as a second control
point but such control reduces again to control by regulation of
cAMP levels. What our laboratory is interested in is the possibility of "post-cAMP" control.

On the one hand, one could argue that there must be such
post-cAMP control mechanisms since without them there would be
an unfocused modification of a broad array of systems. The
whole switchboard would light up. On the surface of it, that
seems imprecise, inelegant and unlikely. On the other hand,
one could argue that there is ample physiologic precedent for
apparently unfocused control mechanisms. Thus, the autonomic
discharge accompanying emotional arousal in mammals is heterogeneous and leads to a host of physiologic responses not all of

which are always exactly appropriate. Yet this example of unfocused response evidently has survival value and has been select-
ed for. By analogy, reduced to the intracellular level, a shot-
gun burst of enzyme changes, while puzzling to us in our present
stage of knowledge, may indeed have survival value—and survival value is all that counts in biological selection. Our long-
term goal in studying the adipocyte is to seek responses linked to
the second messenger but possibly more finely regulated at a third
level of control.

Today I would like to review quickly our evidence that HSL
is activated by cAMP-dependent protein kinase and that that activation is accompanied by phosphorylation. Then, I would like
to present some recent results of our own and of others suggesting
that the effects of insulin in adipocytes are not adequately explained exclusively in terms of alterations of cAMP levels. Our
work at UCSD has been a joint effort with Drs. Jussi Huttunen,
Renu Heller, John C. Khoo, Ms. Alegria Aquino, Ms. Wendy
Fong, Mr. Ray C. Pittman and Mr. Eric Golanty in the Division of Metabolic Disease; Dr. Steven E. Mayer and Mrs.
Barbara Thompson in the Division of Pharmacology; Dr. Leonard
Jarett of Washington University.

LIPASE ACTIVATION BY PROTEIN KINASE

HSL has been purified approximately 100-fold from the
100,000 x g supernatant fraction of rat adipose tissue (20,21).
The purified preparation is a large, lipid-rich particle banding in
sucrose density gradients at about 1.08. It emerges in the void-
volume from 4% agarose columns and in the analytical centrifuge
sediments as a single component with $S_{w,20}$ of 33 to 35*. Negatively stained electron micrographs, prepared by Dr. Thomas
Roth, show a basic particle with diameter approximately 150-200
$\overset{\circ}{A}$ (Fig. 1). Many preparations, however, also show much larger
membrane-like particles, as shown in Fig. 2. Since the purified
preparation contains about 50% by weight phospholipid (Table I)

* We are indebted to Mr. Johannes Everse, Department of Chemistry, for his help in the analytical ultracentrifuge studies.

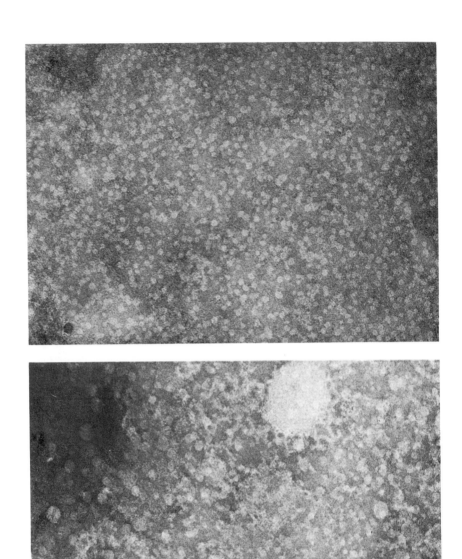

Fig. 1 (above) and Fig. 2 (below). Electron micrographs
of 100-fold purified preparation. Grid-dried sample negatively
stained with 2% phosphotungstate (x 100,000).

TABLE I

Composition of 100-fold purified lipase

Preparation	Protein	Phospholipid (% by weight)	Cholesterol	Triglycerides
No. 22	49	44	7	tr.
No. 23	48	46	6	tr.

we believe the large particles represent aggregates analogous to those demonstrated by Chuang and coworkers to result from recombination of delipidated cytochrome oxidase and phospholipid (22). The purified HSL retains high activity against monoglycerides and diglycerides but, as shown below, there is reason to believe that these activities may not be attributable to the hormone-sensitive triglyceride lipase itself.

Using the 100-fold purified enzyme it was shown that HSL activation (50 to 100%) could be effected by brief incubation with cAMP, protein kinase (from rabbit skeletal muscle), ATP and Mg^{2+} (12). All four additions were essential. The apparent K_m for cAMP was 1.1×10^{-7} M and for ATP 5×10^{-6}. Nucleoside triphosphates other than ATP were ineffective. Cyclic IMP could substitute for cAMP but only at concentrations an order magnitude greater, while other cyclic nucleotides yielded effects only at concentrations so high that contamination with cAMP could not be ruled out as the basis.

Using cruder fractions of adipose tissue, activation can be effected by addition of cAMP, ATP and Mg^{2+} alone, i.e. addition of exogenous protein kinase is not necessary (12,19,23,24). This is attributable to the presence of adequate levels of endogenous protein kinase in such fractions; Corbin and Krebs have demonstrated the presence of high levels of protein kinase in adipose tissue (25). Addition of a specific protein kinase inhibitor blocks such cAMP-stimulated activation and addition of excess exogenous protein kinase restores the activation (10,25,26). The activation reported earlier by Rizack (23) and by Tsai, Belfrage and Vaughan (24) also presumably reflects the presence of protein kinase in their preparations.

PHOSPHORYLATION OF PARTIALLY PURIFIED LIPASE

Using the 100-fold purified HSL, we were able to show that during activation there was transfer of radioactivity from $[\gamma - ^{32}P]ATP$ to enzyme protein (11,12). The time course for activation and phosphorylation were parallel (Fig. 3). The maximum transfer of phosphate to enzyme in 4 such studies corresponded to 2 to 4 moles per 10^6 grams of protein. None of the

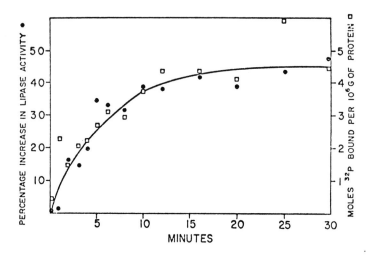

Fig. 3. Time course for activation and phosphorylation of purified hormone–sensitive lipase.

label was associated with the lipid moiety of the preparation. The protein-bound ^{32}P was stable in the presence of 0.25 N HCl but 80% was released in 5 hours at 37° in 0.25 N NaOH, compatible with the postulate that it was bound in O-phosphate linkage. No specific identification of the binding site is available at this time but by analogy with other protein kinase regulated enzymes a serine or threonine acceptor site seems likely.

LOWER GLYCERIDASE ACTIVITIES

As mentioned above, the purified HSL particle retains considerable lipolytic activity against mono- and diglycerides (4). However, these activities are not significantly enhanced by the protein kinase preincubation that consistently enhances triglyceride lipase activity (Table II).

TABLE II

Protein kinase effects on tri-, di- and monoglyceride hydrolase activity in partially purified lipase

Substrate	Activation by complete system (%)
Triolein	70 ± 8
Diolein	15 ± 5
Monoolein	4 ± 2

The monoglyceride hydrolase activity has been more fully characterized and shown to be different in a number of other respects from the triglyceride hydrolase activity. These include different profiles of thermal inactivation; differential inhibition by NaCl, deoxycholate and isopropanol; and differences in pH-activity profile. Thus, there is reason to believe that this large particle may indeed represent a multi-enzyme complex. However, this must remain speculative until physical resolution can be effected.

ACTIVATION OF HUMAN HSL

Thus far protein kinase activation of HSL has been reported only in rat adipose tissue. We have studied crude extracts of adipose tissue from several other species (bovine, ovine, canine)

and, using the conditions that work in rat tissue, have found only small and inconsistent effects. However, Dr. Khoo has recently shown that human adipose tissue HSL behaves very much like that of the rat (26). Absolute levels of lipase activity are of course much lower than in rat tissue but by concentrating the enzyme by isoelectric precipitation from a 100,000 x g supernatant fraction it is quite feasible to work with the amounts of tissue available as surgical biopsies at the time of laparotomy. Activation in this relatively crude preparation does not depend on addition of exogenous protein kinase. Addition of protein kinase inhibitor completely blocks activation, however, and addition of progressively larger amounts of exogenous protein kinase (from rabbit muscle) restores activation (Table III). The difficulty in demonstrating activation in adipose tissue of other species may only reflect different optimal conditions but the possibility that alternate mechanisms for activation are operative cannot be ruled out.

TABLE III

Block of activation of human HSL
by protein kinase inhibitor (PKI) and its
restoration by addition of excess protein kinase (PK)

Additions	Percentage activation
ATP (0.5 mM) + cAMP (0.01 mM)	88
ATP, cAMP, PKI (7.8 μg)	0
As above + 12.3 μg PK	14
" + 24.5 μg PK	40
" + 36.8 μg PK	78
" + 32.5 μg PK	71

50 μl of isoelectric precipitate fraction (in 0.25 M sucrose, 1 mM EDTA) in final volume of 200 μl. Activation 10 min at pH 8.0; assay at pH 6.8 against triolein-^{14}C 1 hr at 30°.

THE EFFECTS OF INSULIN

Recently Khoo et al. (10) have studied simultaneously the

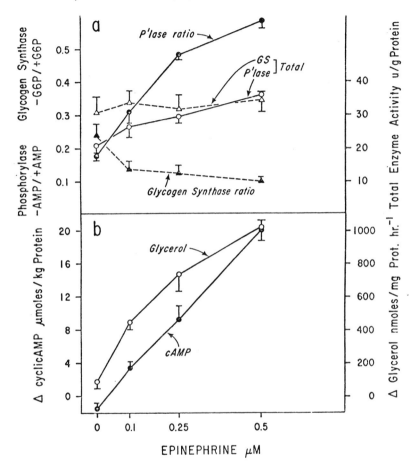

Fig. 4. Dose-response curves relating changes in cAMP accumulation, glycerol release, glycogen synthase and phosphorylase activity in rat adipocytes incubated for 5 min in the presence of the indicated concentrations of epinephrine. Phosphorylase activity measured in the presence or absence of 2 mM 5'-AMP; glycogen synthase activity measured in the absence or presence of 7.5 mM glucose 6-phosphate. Each point represents the mean (± SEM) of results from four experiments.

responses to epinephrine of HSL, phosphorylase and glycogen synthase in rat adipocytes and shown a close parallelism between changes in cAMP production and changes in the activity of these three enzymes, as indicated by the dose-response curves in Fig. 4. Insulin, in the presence of 10 mM glucose, effectively blocked the epinephrine effects on all three systems (Table IV). However,

TABLE IV

Effects of epinephrine (0.5 µM) and insulin (50 µU/ml)
on adipocyte phosphorylase, glycogen synthase, glycerol
release and cAMP accumulation in presence of 10 mM glucose

Addition	Phosphorylase activity ratio*	Glycogen synthase ratio*	Glycerol release** $\left(\dfrac{\mu moles/hr}{mg\ protein}\right)$	cAMP production $\left(\dfrac{pmoles/5\ min}{mg\ protein}\right)$
None	0.19 + 0.02	0.23 + 0.03	0.16 + 0.04	10.0 + 1.9
Epinephrine	0.50 + 0.06	0.13 + 0.01	1.10 + 0.14	26.8 + 5.8
Epinephrine plus insulin	0.24 + 0.04	0.22 + 0.03	0.22 + 0.02	20.1 + 3.4

* As in legend to Fig. 4.
** Measured over 5 min but expressed as hourly rate.

the production of cAMP was only minimally inhibited under the same conditions. The cAMP production in the presence of both epinephrine and insulin was still at least twice that in control incubations and yet the rates of glycerol production and the values for phosphorylase activity ratio and glycogen synthase ratio were little different from control values. The values for cAMP production in the presence of epinephrine plus insulin are not significantly different from those in the presence of epinephrine alone when analyzed as grouped data. Analyzed as paired data, however, the difference is statistically significant but it is small. The essential point is that cAMP levels like those seen in the presence of both hormones are associated with much higher rates of glycerol release and larger changes in target enzymes when reached in the presence of epinephrine alone.

These results make it difficult to attribute the insulin effect directly and exclusively to its ability to modulate cAMP levels, whether through effects on adenylate cyclase or on phosphodiesterase. Under some conditions insulin does indeed have clearcut effects on hormone-induced increases in cAMP. Thus, Butcher et al. (27) showed that in the presence of caffeine insulin sharply reduced the epinephrine-induced increases in cAMP level. However, the cAMP levels reached in the presence of both insulin and epinephrine were still much higher than in control cells and glycerol production was not significantly reduced. Moreover, in other experiments glycerol production was inhibited without significant suppression of cAMP production, as in the present studies. Jarett et al. found that under the conditions they used insulin inhibited epinephrine-stimulated glycerol production in rat adipocytes without significantly affecting cAMP levels reached (28). They proposed that compartmentalization of cAMP might account for the apparent paradox, concentration in some "relevant" subcompartment being in fact reduced by insulin. Burns, Langley and Robison have reported a similar dissociation in human adipocytes (29).

Dissociation between insulin-epinephrine antagonism and effects on cAMP levels have been noted in other tissues. Larner and coworkers have shown that insulin blocks the effect of epinephrine on glycogen synthase in skeletal muscle without an effect on cAMP levels (30,31). Walaas and Walaas reported similar results with regard to changes in phosphorylase activity in diaphragm (32). They attributed the insulin effect to changes in protein kinase activity not necessarily affected by changes in cAMP concentration per se.

At least four potential sites of insulin-epinephrine antagonism can be visualized (Fig. 5). Action at site A exclusively seems unlikely in view of the evidence cited above. Action at site D seems improbable as it would require invoking additional parallelisms in three quite different enzyme systems. Moreover, in the case of at least one of these the activation is indirect (phosphorylase activation via phosphorylase kinase). Action at site C—general interference with the interaction of protein kinase

57

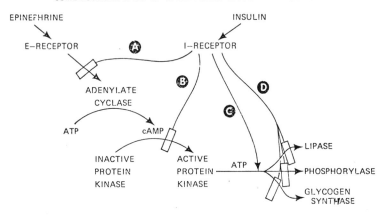

SOME POSSIBLE SITES OF EPINEPHRINE—INSULIN ANTAGONISM

Fig. 5. Some possible sites at which epinephrine-insulin antagonism might be exercised.

and its substrates — is compatible with the observed results. However, insulin itself is unlikely to be the moderator. We have added insulin to the complete system of purified lipase and observed no inhibition (unpublished results). Action at site B could explain all of the observed effects and, most important, could rationalize the effects on expressed protein kinase activity without "appropriate" concomitant changes in cAMP levels. An effect of insulin on synthase kinase (protein kinase) in diaphragm has been reported by Villar-Palasi and coworkers (33,34) and by Miller and Larner (35). Also, Soderling and coworkers (36) have observed shifts in the ratio of cAMP-dependent to cAMP-independent protein kinase in fat tissue induced by epinephrine and opposed by insulin. Coupled with data showing changes in cAMP levels too small to account for such shifts this would support action at site B. Of course, compartmentalization remains a possibility but the evidence available strongly suggests an effect of insulin exerted in part at least beyond cAMP, i.e. at a "third level".

REFERENCES

(1) M. Vaughan, J. Berger and D. Steinberg. J. Biol. Chem. 239 (1964) 401.

(2) O. Strand, M. Vaughan and D. Steinberg. J. Lipid Res. 5 (1964) 554.

(3) D. Steinberg, in: Progress in Biochemical Pharmacology, Vol. 3, eds. D. Kritchevsky, R. Paoletti and D. Steinberg (Karger, Basel, 1967) p. 139.

(4) R.A. Heller and D. Steinberg. Biochim. Biophys. Acta 270 (1972) 65.

(5) M. Vaughan, D. Steinberg and E. Shafrir. J. Clin. Invest. 38 (1959) 1051.

(6) M. Vaughan. J. Biol. Chem. 235 (1960) 3049.

(7) D. Steinberg, M. Vaughan, P. Nestel, O. Strand and S. Bergström. J. Clin. Invest. 43 (1964) 1533.

(8) R.L. Jungas. Proc. Nat. Acad. Sci. (USA) 56 (1966) 757.

(9) J. Moskowitz and J.N. Fain. J. Clin. Invest. 48 (1969) 1802.

(10) J.C. Khoo, D. Steinberg, B. Thompson and S.E. Mayer. Submitted for publication (1972).

(11) J.K. Huttunen, D. Steinberg and S.E. Mayer. Biochem. Biophys. Res. Commun. 41 (1970) 1350.

(12) J.K. Huttunen and D. Steinberg. Biochim. Biophys. Acta 239 (1971) 411.

(13) R.J. DeLange, R.G. Kemp, W.D. Riley, R.A. Cooper and E.G. Krebs. J. Biol. Chem. 243 (1968) 2200.

(14) S. Hizukuri and J. Larner. Biochemistry 3 (1964) 1783.

(15) G.I. Drummond and L. Duncan. J. Biol. Chem. 241 (1966) 5893.

(16) E.G. Krebs. Curr. Top. Cell. Regulat. 5 (1972) 99.

(17) E.H. Fischer, L.M.G. Heilmeyer, Jr. and R.H. Haschke. Curr. Top. Cell. Regulat. 3 (1971) 211.

(18) J. Larner and C. Villar-Palasi. Curr. Top. Cell. Regulat. 4 (1971) 195.

(19) J.D. Corbin, E.M. Reiman, D.A. Walsh and E.G. Krebs. J. Biol. Chem. 245 (1970) 4849.

(20) J.K. Huttunen, J. Ellingboe, R.C. Pittman and D. Steinberg. Biochim. Biophys. Acta 218 (1970) 333.

(21) J.K. Huttunen, A.A. Aquino and D. Steinberg. Biochim. Biophys. Acta 224 (1970) 295.

(22) T.F. Chuang, F.F. Sun and F.L. Crane. Bioenergetics 1 (1970) 227.

(23) M.A. Rizack. J. Biol. Chem. 236 (1961) 657.

(24) S.-C. Tsai, P. Belfrage and M. Vaughan. J. Lipid Res. 11 (1970) 466.

(25) J.D. Corbin and E.G. Krebs. Biochem. Biophys. Res. Commun. 36 (1969) 328.

(26) J.C. Khoo, W.W. Fong and D. Steinberg. Biochem. Biophys. Res. Commun. 49 (1972) 407.

(27) R.W. Butcher, J.G.T. Sneyd, C.R. Park and E.W. Sutherland, Jr. J. Biol. Chem. 241 (1966) 1651.

(28) L. Jarett, A.L. Steiner, R.M. Smith and D.M. Kipnis.
      Endocrinology 90 (1972) 1277.

(29) T.W. Burns, P.E. Langley and G.A. Robison, in: Advances
      in Cyclic Nucleotide Research, Vol. 1, eds. P. Greengard,
      R. Paoletti and G.A. Robison (Raven Press, New York,
      1972) p. 63.

(30) N.D. Goldberg, C. Villar-Palasi, H. Sasko and J.
      Larner. Biochim. Biophys. Acta 148 (1967) 665.

(31) J.W. Craig, T.W. Rall and J. Larner. Biochim. Biophys.
      Acta 177 (1969) 213.

(32) O. Walaas and E. Walaas, in: Advances in Cyclic Nucleo-
      tide Research, Vol. 1, eds. P. Greengard, R. Paoletti and
      G.A. Robison (Raven Press, New York, 1972) p. 590.

(33) C. Villar-Palasi and J.I. Wenger. Fed. Proc. 26 (1971)
      563.

(34) L.C. Shen, C. Villar-Palasi and J. Larner. Physiol.
      Chem. Phys. 2 (1970) 536.

(35) T.B. Miller and J. Larner. Proc. Nat. Acad. Sci. (USA)
      69 (1972) 2774.

(36) T.R. Soderling, J.D. Corbin and C.R. Park. Fed. Proc.
      31 (1972) 440 Abs.

DISCUSSION

C. DALTON: I am curious about your rationale for including theophylline in your activation conditions for the isolated enzyme. Is the phosphodiesterase still present in this preparation? What happens if you do not include the theophyline?

D. STEINBERG: It is not needed. It was included originally, but it does not improve the activation and it can be omitted even from fairly crude preparations. There is no phosphodiesterase activity in the purified preparation. In more recent experiments we did not use theophyline.

G.S. LEVEY: In your scheme you did not mention insulin effects on phosphodiesterase. Loten and Sneyd showed that in adipose tissue, insulin activated the phosphodiesterase. Would you comment on that report? Secondly, if insulin affects the protein kinase directly, would insulin have to enter the cell? Do you have any data on this point?

D. STEINBERG: I'm aware of the paper by Loten and Sneyd on the effect on phosphodiesterase. As I indicated earlier, it doesn't matter whether one is talking about an effect on phosphodiesterase to increase its activity or a suppression of adenylate cyclase activity. The paradox is that the cyclic AMP levels do not seem to reflect a change in either of those. In other words I do not think that the phosphodiesterase effect postulated, (although not confirmed by everybody), gets us off the hook on the issue of why the cyclic AMP levels do not seem to change in a manner adequate to account for the changes in the regulated enzymes. I certainly do not know whether insulin has to get in or not. We have tried some mixing experiments to see if we could get evidence for something formed in an insulin-treated cell such that its homogenate could influence activation in the second homogenate, but I can not say that we have found anything exciting yet.

J.D. CORBIN: I wonder if you've determined whether activation of the lipase by protein kinase changes the K_m or the V_{max} of the enzyme for triglyceride?

D. STEINBERG: We have looked at that, and find no difference so far. That is to say, saturation curves for substrate look the same whether we use enzyme prepared from hormone-treated tissue or whether we take enzyme from control tissue. In neither case do we have, I think, exclusively activated or exclusively non-activated enzyme to compare, so I don't think a negative result in the absence of such a resolution rules out the possibility of a difference.

K.S. SIDHU: Would you speculate on what inactivates the activated hormone-sensitive lipase?

D. STEINBERG: The dogma of course requires that it be a phosphatase, but we can not do any more than speculate. If we incubate homogenates at 37°, and particularly at an elevated pH, the lipase is inactivated by about 50% in 20 minutes. But then when we add the protein kinase cyclic AMP/ATP system to it, we cannot get it back up to the original level. Dr. Vaughan and Dr. Tsai have reported an ATP-dependent inactivation of lipase that depends upon some small molecular weight component not yet identified. This ATP-magnesium dependent inactivation was not reversible. The difficulty may lie in the presence of some other inactivation systems, but at the moment I do not think we can say more than that we are trying to demonstrate phosphatase inactivation of the lipase, but we have not succeeded.

J. FESSENDEN-RADEN: I have two questions on the lipase. The first is, are both the cholesterol and the phospholipid required for lipase activity and two, if you reduce the lipid content in the lipase or even remove the lipid entirely, can the lipase still be phosphorylated?

D. STEINBERG: The extraction with organic solvents to the point of total removal of the lipid leaves us with a totally inactive enzyme. We never tried the phosphorylation. That is a very good idea. In fact, from what Dr. Krebs has told us it might be a better substrate than the original lipase. We should possibly go back and try that. You can extract gently (and therefore not remove a large fraction of the lipid) and still have activity. Recombination experiments in which, the extracted lipid is added back to the protein, does not restore lipase activity. SDS treatment totally destroys the activity.

J. FESSENDEN-RADEN: Is this both with the cholesterol and the phospholipid and have you kept the ratio of phopholipid to cholesterol constant.?

D. STEINBERG: We have not analysed what fraction of which lipid was being removed.

E.G. KREBS: I have a medical student-type question. What good does it do for the fat cell to break down glycogen at the same time that it releases fatty acids?

D. STEINBERG: That's one of those deep questions. I have a three page speculation about that, but no answers. There is reesterification at a brisk rate when you stimulate with hormones. Martha Vaughan and I showed, using a balance method so that you can estimate both synthesis and degradation, that when you stimulate with ACTH or epinephrine the rate of reesterification may increase two- to three-fold from basal levels, while of course, lipolytic activity increases to a larger extent so you get net lipolysis. If you did not have available α-glycerol phosphate, you could not sustain that rate of reesterification and might get a buildup of free fatty acids damaging to the cell. There are other possibilities too, such that the breakdown of glycogen may in some way relate to making lipid available for lipolysis. We may speculate about how this huge lipid droplet with relatively limited surface area compared to its weight, becomes available as a substrate during hormone action. We might draw little pictures of glycogen possibly sitting around that lipid droplet. The removal of the glycogen rather that generation of the glycolytic intermediates may be the reason why glycogenolysis is coincident with lipolysis.

H. SHEPPARD: I am somewhat bothered by the implication of your results that a forty to one hundred percent increase in the lipase activity could account for a greater than ten fold increase in lipolysis in the cell.

D. STEINBERG: This is, of course, one of the things that bother us also. With isolated adipolytes, which start with essentially zero lipolytic activity, you can get as high as twenty-fold, thirty-fold increases in rate of glycerol release and yet we only seem to get at best a two-fold increase in lipase activity. That has caused us to look for effectors that may differentially alter the activity of the activated and the non-activated forms. This might be responsible for the enhancement of response in the intact system. Second, we wonder if we are getting some artifactual activation during the preparation of the control tissue. Third, it may be that the presentation of substrates that we use (gum arabic-stabilized triolein emulsions) is inappropriate and that with presentation in a more physiologic form one would see the larger difference one would like to see between activated and non-activated enzyme.

H. SHEPPARD: Is it possible that one is dealing with a form of the enzyme which is either associated or not associated with the fat droplet? If so, the activity per se wouldn't be as important as its distribution in the cell.

D. STEINBERG: Yes, this is equally possible. Compartmentalization effects are possible.

MOLECULAR CHARACTERIZATION OF CYCLIC AMP-DEPENDENT PROTEIN KINASES DERIVED FROM BOVINE HEART AND HUMAN ERYTHROCYTES

O.M. ROSEN, C.S. RUBIN and J. ERLICHMAN
Departments of Medicine and Molecular Biology
Albert Einstein College of Medicine

INTRODUCTION

For several years our laboratory has been studying some of the enzymes involved in the synthesis and degradation of cyclic 3',5'-adenosine monophosphate (cyclic AMP). After Krebs and his colleagues discovered that cyclic AMP regulated the process of glycogenolysis by activating phosphorylase kinase kinase and that the latter enzyme was able to catalyze the transfer of phosphoryl groups from ATP to a number of protein substrates (1) we initiated a study of the cyclic AMP-dependent protein kinase of bovine heart. We planned to purify the protein, analyze its molecular structure and investigate the mechanism of activation by cyclic AMP. During the course of this work, the effects of cyclic AMP on a variety of membrane-related phenomena such as slime mold aggregation (2) contact inhibition (3) and secretion (4) were uncovered. These developments coupled with the proposal (5) that the activation of protein kinases by cyclic AMP is the final common pathway for cyclic nucleotide action in eukaryotic cells, led us to ask whether cyclic AMP-dependent protein kinases were associated with cell membranes and, if so, to what extent the structure and function of kinases could be influenced by this association. In this article we will summarize the results of studies carried out on the soluble cyclic AMP-dependent protein kinase of bovine heart as well as the membrane-associated cyclic AMP-dependent protein kinase of human erythrocytes.

PROTEIN KINASE OF BOVINE HEART

In the initial attempts to purify the cyclic AMP-dependent protein kinase, fresh beef hearts were homogenized and the soluble kinase was fractionated with ammonium sulfate, selectively eluted from calcium phosphate gel and filtered on Sephadex G-200 (6). During the latter step, approximately 90% of the protein kinase activity emerged as a single peak coincident with 90% of the cyclic AMP-binding activity. Proteins which possessed only cyclic AMP-binding or protein kinase activities were not observed. The principal cyclic AMP-dependent protein kinase of bovine heart was estimated to have a molecular weight of 240,000 utilizing the technique of gel filtration according to the method of Andrews (7). Studies to be discussed subsequently, using highly purified enzyme and filtration on Bio-Gel P-300 also suggested that the enzyme had a molecular weight in the range of 240,000-280,000. The enzyme recovered after filtration on Sephadex G-200 (100-fold purified) was then adsorbed to DEAE-cellulose which had been equilibrated with 0.05M potassium phosphate buffer, pH 7.0. In an effort to take advantage of the high affinity of protein kinases for cyclic AMP, 1μM cyclic AMP was added to the equilibrating buffer and applied to the column after adsorption of the enzyme. Happily, more than 80% of the applied kinase units were eluted resulting in a substantial purification. The elution step was specific for cyclic AMP; 1μM cyclic GMP was approximately 25% as effective as cyclic AMP and 5'-AMP was ineffective. Some of the properties of the eluted kinase were, however, unlike those of the enzyme in crude extracts or in partially purified preparations prior to the DEAE step. Whereas phosphorylation of protamine by the original enzyme was stimulated 3-4-fold by 1μM cyclic AMP (K_a for cyclic AMP: 6.2×10^{-8}M), the DEAE-purified enzyme, even after exhaustive dialysis, was neither stimulated by nor capable of binding cyclic nucleotides. Following purification on DEAE, the enzyme was also unstable, rapidly inactivating upon freezing or heating at 45° for 10 minutes. Properties shared by the eluted enzyme and the less purified preparation included a K_m for ATP of 13-14μM, the same rates for phosphorylation of protamine, histone and casein (in the presence of cyclic AMP), and identical divalent metal requirements and pH optima (7.8). When applied to Sephadex G-200, the cyclic AMP-independent protein

kinase was significantly more retarded than the native cyclic AMP-dependent kinase and appeared to have a molecular
weight of approximately 30,000-40,000. The cyclic AMP-
binding activity which was not eluted from DEAE with the
kinase could be recovered from the resin (without further
purification) by elution with high concentrations of salt.
As would be anticipated from the good recovery of kinase
units during the cyclic AMP-elution step, this cyclic nucleotide-binding activity was not associated with kinase
activity. When the fraction containing binding activity
was subjected to gel filtration, a peak of cyclic AMP-binding activity emerged in an elution volume corresponding to
a molecular weight of about 160,000. It thus appeared, as
had been suggested for the cyclic AMP-dependent protein
kinases of the adrenal cortex (8), reticulocyte (9), muscle
(10), and liver (11), that the beef heart protein kinase
could be dissociated by cyclic AMP into smaller, dissimilar
units: a cyclic AMP-independent protein kinase and a cyclic
AMP-binding moiety. If the mechanism of cyclic AMP activation involves dissociation of the protein kinase into its
two components and if this reaction is reversible, reassociation into a cyclic AMP-dependent protein kinase should be
possible. Aliquots containing cyclic AMP-independent kinase
and cyclic AMP-binding protein were consequently combined,
dialyzed together against phosphate buffer and assayed for
diminution in protein kinase activity which could be over-
come by the addition of cyclic AMP. Evidence for restoration of cyclic AMP-dependency was obtained after dialysis at
23° but not after dialysis at 4° suggesting that the dissociation of bound cyclic nucleotide, subunit reassociation,
or both, occur more effectively at the higher temperature.
Documentation that restoration of cyclic nucleotide activat-
ability is accompanied by physical recombination of the two
kinds of subunits was then provided by showing that upon gel
filtration of the reformed cyclic AMP-dependent enzyme, cyclic AMP-binding and protein kinase activities emerged in
the same elution volume (6).

A new procedure for purification of the bovine heart enzyme which does not involve exposure to cyclic AMP, dissociation and reassociation, was then devised (See Table 1).
This procedure, employing standard techniques for protein
separation, resulted in a 1,200-fold purification (12) of
both the cyclic AMP-binding and protein kinase activities.

TABLE 1

Purification of beef heart protein kinase (12)

Step	Protein mg	Units		Specific activity	
		Catalytic $\times 10^{-3}$	Binding	Catalytic	Binding
Homogenate (10,000 X g supernatant)	298,250	193.9	2326	0.65	0.0078
Ammonium sulfate	87,954	192.8	6245	2.2	0.071
DEAE-Sephadex	10,585	142.2	2752	14	0.26
Alumina Cγ	2,012	98.6	1730	49	0.86
DEAE-cellulose	510	57.1	1071	112	2.1
Bio-Gel P-300	111	31.9	488	287	4.4
Hydroxylapatite	37	29.9	333	807	9.0

Throughout the purification, the enzyme retained its responsiveness to cyclic nucleotides and its kinetic properties remained unaltered. The purified enzyme appeared homogeneous when analyzed by polyacrylamide disc gel electrophoresis. Upon storage at 4°, however, two species of cyclic AMP-dependent protein kinase activities were generated
which could be isolated by refiltration on Bio-Gel P-300.
All of the protein components resolved by gel filtration or
electrophoresis contained cyclic nucleotide-dependent protein kinase of similar specific activity but the proportions

of the minor components increased with duration of storage[1].
Curiously, the molecular weight of the principal protein
kinase species, estimated to be 240,000-280,000 by gel filtration, appeared to have a molecular weight of only 140,000
-170,000 when estimated by the electrophoretic method of
Hedrick and Smith (13) or by techniques of sedimentation
velocity and sedimentation equilibrium. For this reason we
have recently reexamined the molecular parameters of the
enzyme (See Table 2) utilizing gel filtration data for the
calculation of Stokes radius and combining this data with
the sedimentation constant obtained from sucrose gradient
ultracentrifugation. The data suggest that the cyclic AMP-
dependent form of the enzyme may be asymmetric (accounting
for its high apparent molecular weight on gel filtration)
and that the correct value for its molecular weight is
closer to the values estimated by gel electrophoresis and
ultracentrifugation, i.e. 174,000.

TABLE 2

Physical properties of beef heart protein kinase

Form	Stokes radius (Å)	Sedimen- tation constant	Friction- al ratio	Axial ratio	Molecular weight (X 10^{-3})
cAMP-Dependent Kinase (K)	60	6.8	1.62	12	174
cAMP-Binding Protein (B)	50	4.6	1.64	12	98
		(SDS electrophoresis:			55)
cAMP-Independ- ent Kinase (C)	25	3.6	1.12	3	38
		(SDS electrophoresis:			42)

K = 2C + 1B (dimer)

----

[1] This development of heterogeneity can be minimized by
storage of the enzyme at 4° in 85% saturated ammonium sulfate containing 2 mM EDTA.

The subunit composition of the purified cyclic AMP-dependent protein kinase (freshly prepared or stored) was then studied by denaturing the enzyme in SDS and subjecting it to SDS-acrylamide gel electrophoresis. Two polypeptide chains with molecular weights of 42,000 and 55,000 were resolved. Homogeneous, cyclic AMP-independent protein kinase was then prepared from the purified cyclic AMP-dependent protein kinase by the DEAE-cyclic AMP procedure previously described. After elution of the kinase activity from DEAE, the cyclic AMP-binding protein was recovered in homogeneous form by elution with 0.25M potassium phosphate buffer, pH 7.0. When these two proteins were subjected to SDS-electrophoresis, the cyclic AMP-binding protein gave one band which corresponded to the larger component of the cyclic AMP-dependent protein kinase. The cyclic AMP-independent protein kinase also yielded a single protein band corresponding to the smaller of the original two components. Thus the species of molecular weight 42,000 could be designated the cyclic AMP-independent kinase or catalytic component (C) and the component of molecular weight 55,000, the cyclic AMP-binding protein (B). When the cyclic AMP-independent kinase and the cyclic AMP-binding moieties were subjected to the same kind of physical analysis as the native, undissociated enzyme, molecular weights of 38,000 and 98,000, respectively, were derived (Table 2). It is interesting to note that the frictional ratios of the binding protein and the intact kinase are identical and significantly greater than 1.2 suggesting that both are asymmetric molecules. The catalytic moiety on the other hand appears to be globular. Our current working hypothesis for the molecular structure of the cyclic AMP-dependent protein kinase of bovine heart is that it is asymmetric, has a molecular weight of 174,000 and is composed of two globular catalytic subunits (molecular weight 38,000 each) and an asymmetric cyclic AMP-binding protein which exists as a dimer of molecular weight 98,000. From studies of cyclic AMP-binding, it appears that 0.9 moles of cyclic AMP are bound per 174,000 grams of protein.

Although cyclic AMP activates protein kinases in vitro by inducing their dissociation, it is possible that in vivo, an active "cyclic AMP-protein kinase" complex may exist and that other factors such as the ability of protein substrates to alter the quaternary structure of the enzyme (14) or the activities of phosphoprotein phosphatases (15), inhibitory

proteins (16,17), or other nucleotides (18,19) may play significant roles in regulating protein kinase activity. Recently, we have found that purified protein kinase from
bovine heart can utilize the γ- phosphate of ATP to catalyze its own phosphorylation. Analysis of the product of
this phosphotransferase reaction indicates that it has the
properties of an ester phosphate of serine or threonine
(hot acid and hydroxylamine stable, alkali labile) and that
the protein substrate is the cyclic AMP-binding component of
the protein kinase. Phosphorylation occurs rapidly in the
presence or absence of cyclic AMP and 0.9 mole phosphate is
incorporated per cyclic AMP-binding peptide chain of molecular weight 55,000. There is, as yet, no evidence that phosphorylation of protein kinase has physiological significance
but experiments to assess this are underway. Another observation which supports the possibility of a multiplicity of
controls of protein kinase activity is the stimulation by
polyarginine (100μg/ml) of the basal and cyclic AMP-stimu-
lated activities of beef heart protein kinase on certain
protein substrates. Thus the addition of polyarginine
stimulates the phosphorylation of bovine serum albumin
(a relatively poor subst;ate for this enzyme) four-fold in
the absence of added cyclic AMP and two-fold in the presence
of optimal concentrations of cyclic AMP. On the other hand,
the phosphorylation of protamine is totally unaffected by
the addition of polyarginine.

PROTEIN KINASE OF HUMAN ERYTHROCYTE MEMBRANES

Mature mammalian erythrocytes are a good source of com-
paratively well characterized plasma membranes and since
they are also a readily accessible human tissue, it seemed
reasonable to determine whether human erythrocytes contained
a membrane-bound protein kinase. Although mature mammalian
erythrocytes have not been assigned any cyclic AMP-mediated
function, they have been shown to contain adenylate cyclase
and cyclic nucleotide phosphodiesterase activities (20).
Approximately 70% of the total human erythrocyte protein
kinase activity, assayed in a standard reaction mixture
containing protamine was found to be membrane associated
(21). The rate of phosphorylation of protamine was enhanced 5-8-fold by the addition of 2μM cyclic AMP. In the
absence of added exogenous protein acceptors, phosphorylation of endogenous protein acceptors became apparent but the

73

overall rate of this phosphorylation was stimulated only
20-30% by the addition of cyclic AMP. Some of the biochemical properties of the cyclic AMP-dependent protein kinase
of the human erythrocyte membranes are very similar to those
described for the soluble kinases. Casein, histone, and
protamine are good phosphate acceptors but cyclic AMP enhances the phosphorylation of only the latter two. The K_a
for cyclic AMP and the K_m's for protamine, ATP and Mg^{2+}
are similar to those of the soluble beef heart protein
kinase (See Table 3), as are the sensitivities to stimulation by the various cyclic nucleotides (cyclic AMP >
cyclic IMP > cyclic GMP).

TABLE 3

Comparison of beef heart and human erythrocyte
protein kinases

	Beef heart	Human erythrocyte
Predominant form	Soluble	Membrane-bound
Protein substrates		
exogenous	Protamine ...	Protamine ...
endogenous	B	3 membrane components
Activation by cAMP	+ (K_a 62nM)	+ (K_a 28nM)
Dissociation into B&C	+ (cAMP)	+ (Salt)
B K_d	22nM	3.3nM
MW	98,000	ND
C K_m (ATP)	13μM	8.3μM
MW	40,000	40,000
Reassociation	+	ND

B, cAMP-binding protein; C, cAMP – independent kinase;
ND, not determined; MW, molecular weight;
K_d is for the cAMP –protein kinase complex

Membranes, like the cyclic AMP-binding component of soluble protein kinases, bind cyclic AMP with an extremely high affinity. Approximately 15 picomoles of cyclic AMP are bound per mg membrane protein. The binding is specific for cyclic AMP and bound cyclic AMP can be released from the membranes by boiling and identified as unaltered nucleotide. As with the soluble cyclic AMP-binding protein (22,23) the ^{3}H-cyclic AMP bound to erythrocyte membranes is susceptible to neither hydrolysis by added cyclic nucleotide phosphodiesterase nor ready displacement by the subsequent addition of non-radioactive cyclic AMP.

One of the characteristics of soluble cyclic AMP-dependent protein kinases is that they can be dissociated into two functionally dissimilar components, one of which binds cyclic nucleotides and the other of which catalyzes the cyclic nucleotide-independent phosphotransferase reaction. When erythrocyte membranes are extracted for two hours at 4° in the presence of salt solutions of high ionic strength such as 1M ammonium chloride, 50-60% of the protein kinase activity is released from the membrane. The solubilized activity is cyclic AMP-independent and has very little cyclic AMP-binding capacity (See Table 4).

TABLE 4

Distribution of erythrocyte protein kinase (21)

Enzyme source	Protamine addition	Protein kinase activity		Percentage of total activity
		-Cyclic AMP	+2μM Cyclic AMP	
		units/mg protein		
Soluble	+	11.8	15.2	16
	-	1.0	1.0	
Membranes	+	134	734	84
	-	30.8	34.7	

Analysis of the solubilized enzyme by sucrose gradient ultra-centrifugation yields a molecular weight of about 40,000, very similar to the molecular weight of the analogous compon-ent of soluble protein kinases. The particulate residue re-tains all of the original cyclic nucleotide-binding activity. Thus the membrane-associated protein kinase would appear to have many features in common with soluble protein kinases including the ability to dissociate into cyclic AMP-binding and cyclic AMP-independent protein kinase components. In this situation, the catalytic moiety appears to be loosely bound to the membrane whereas the cyclic AMP-binding moiety is more firmly integrated into the membrane structure. Ap-proximately 50% of the erythrocyte membrane protein kinase can also be solubilized by extraction with 0.1mM EDTA at pH 8. Under these conditions, however, dissociation does not occur and the solubilized enzyme retains its ability to bind and to be stimulated by cyclic AMP.

Intact erythrocytes, unlike the purified membranes de-rived from them, do not bind cyclic AMP and efforts to demon-strate phosphorylation of membrane components by the addition of ATP to intact cells were not successful. It would seem that neither the active site of the kinase (&/or its endo-genous substrates) nor the cyclic AMP-binding site are freely exposed on the external surface of the intact erythrocyte. Although these observations make it unlikely that the pro-tein kinase responds to cyclic AMP in the plasma, the finding of some adenylate cyclase activity as well as intracellular cyclic AMP concentrations of approximately 10^{-8}M in mature human erythrocytes, leaves open the possibility that kinase activity may be influenced by endogenous cyclic AMP metabo-lism. Naturally occurring activators of human erythrocyte adenylate cyclase have not yet been identified.

Although protamine is a useful substrate for assaying protein kinase activity, endogenous substrates have to be identified in order to understand the physiologi-cal functions of these enzymes. Accordingly, the membrane components phosphorylated by the membrane-associated protein kinase of the human erythrocyte have been studied (24). Erythrocyte membranes were phosphorylated in the presence or absence of cyclic AMP, dissociated into their component pep-tide chains by treatment with SDS and resolved by SDS-acryl-amide gel electrophoresis. Using the protein band and molecular weight assignments of Fairbanks et al. (25), only

76

three of the protein components of the membrane are phosphorylated. Approximately 60% of the total phosphate incorporated is associated with protein II (molecular weight: 215,000).
This protein is rapidly phosphorylated but the rate of phosphorylation is not enhanced by the addition of cyclic AMP.
Protein II has been estimated to represent approximately 10-
15% of the total erythrocyte membrane protein (25), its
function is unknown. The other two proteins which are phosphorylated are protein III (molecular weight: 88,000) and a
protein which we have designated IVc (molecular weight:
50,000). The rates of phosphorylation of proteins III and
IVc are stimulated 2- and 5-fold, respectively, by the addition of 2 μM cyclic AMP. At present there is some ambiguity
about the phosphorylation assigned to protein III since
under standard conditions of electrophoresis, a glycoprotein
component of erythrocyte membranes comigrates with protein
III. The phosphorylation of protein IVc, a minor (2-4%)
component of the membrane, is greatly enhanced by cyclic AMP.
As with the other two membrane substrates, no conclusive
information is available as to its function. A preliminary
estimate of the stoichiometry of phosphorylation of membranes in vitro is that 0.79 moles phosphate are incorporated
per mole of protein II; 0.024 moles per mole of protein III
and 0.17 moles per mole of protein IVc. The phosphate incorporated into these components is alkali labile as well as
hot acid and hydroxylamine stable. Of the total phosphate
incorporated into the membranes, 70% has been identified as
phosphoserine (59%) and phosphothreonine (11%). When the
total phosphorylation of membranes is measured, the cyclic
AMP -dependent phosphorylation of proteins III and IVc is ob-
scured by the high level of cyclic AMP-independent phosphorylation of protein II. The erythrocyte membrane-associated
protein kinase (or protein kinases) appears able to phosphorylate endogenous and exogenous protein substrates in
both cyclic AMP-dependent and cyclic AMP-independent fashions.

CONCLUSIONS

A comparison of some of the properties of the protein
kinases of beef heart and erythrocyte membranes is presented
in Table 3. In order to understand the function of these
kinases it will be necessary to (a) proceed with an analysis
of their chemical structure and subunit interactions; (b) to
pursue the leads which suggest that the activity of protein

kinases may be regulated by mechanisms in addition to cyclic AMP-induced dissociation, and that cyclic AMP-binding proteins and cyclic AMP-independent protein kinases may have functions other than or in addition to those attributable to the known effects of protein kinases; and (c) to define the endogenous protein substrates for kinase activity and the role that phosphorylation plays in their physiological function.

REFERENCES

(1) D.A. Walsh, J.P. Perkins and E.G. Krebs, J. Biol. Chem. 243 (1968) 3763.

(2) J.T. Bonner, D.L. Barkley, E.M. Hall, T.M. Konijn, J.W. Mason, G.O'Keefe and P.B. Wolfe, Dev. Biol. 20 (1969) 72.

(3) J.Otten, G.S. Johnson, and I. Pastan, Biochem. Biophys. Res. Commun. 44 (1971) 1192.

(4) F. Labrie, S. Lemaire, G. Poirier, G. Pelletier and R. Boucher, J. Biol. Chem. 246 (1971) 7311.

(5) J.F. Kuo and P. Greengard, Proc. Nat. Acad. Sci. U.S.A. 64 (1969) 1349.

(6) J. Erlichman, A.H. Hirsch and O.M. Rosen, Proc. Nat. Acad. Sci. U.S.A. 68 (1971) 731.

(7) P. Andrews, Biochem. J. 91 (1964) 222.

(8) G. Gill and L.D. Garren, Biochem. Biophys. Res. Commun. 39 (1970) 335.

(9) M. Tao, M. Salas and F. Lipmann, Proc. Nat. Acad. Sci. U.S.A. 67 (1970) 408.

(10) M.A. Brostrom, E.M . Reimann, D.A. Walsh and E.G.Krebs, Advan. Enzyme Regul. 8 (1970) 191.

(11) A. Kumon, H. Yamomura and Y. Nishizuka, Biochem. Biophys. Res. Commun. 41 (1970) 1290.

(12) C.S. Rubin, J. Erlichman and O.M. Rosen, J. Biol.Chem. 247 (1972) 36.

(13) J.L. Hedrick and A.J. Smith, Arch. Biochem. Biophys. 126 (1968) 155.

(14) E. Miyamoto, G.L. Petzold, J.S. Harris and P. Greengard, Biochem. Biophys. Res. Commun. 44 (1971) 305.

(15) M.H. Meisler and T.A. Langan, J. Biol. Chem. 244 (1969) 4961.

(16) M.M. Appleman, L. Birnbaumer and H.N. Jones, Arch. Biochem. Biophys. 116 (1966) 39.

(17) D.A. Walsh, C.D. Ashby, C. Gonzalez, D. Calkins, E.H. Fischer and E.G. Krebs, J. Biol. Chem. 246 (1971) 1977.

(18) J.F. Kuo and P. Greengard, J. Biol. Chem. 245 (1970) 2493.

(19) M.K. Haddox, N.E. Newton, D.K. Hortle and N.D. Goldberg, Biochem. Biophys. Res. Commun. 47 (1972) 653.

(20) H. Sheppard and C.R. Burghardt, Molec. Pharmacol. 6 (1970) 425.

(21) C.S. Rubin, J. Erlichman and O.M. Rosen, J. Biol. Chem. 247 (1972) 6135.

(22) R.F. O'Dea, M.K. Haddox and N.D. Goldberg, J. Biol. Chem. 246 (1971) 6183.

(23) A.G. Gilman, Proc. Nat. Acad. Sci. U.S.A. 67 (1970) 305.

(24) C.S. Rubin and O.M. Rosen, Biochem. Biophys. Res. Commun., in press.

(25) G. Fairbanks, T.L. Steck and D.F.H. Wallach, Biochemistry 10 (1971) 2606.

ACKNOWLEDGEMENTS

These investigations were supported by grants from the U.S. Public Health Service and the American Cancer Society.

DISCUSSION

G.H. DIXON: A question one faces with any isolated preparation, even a membrane preparation, is whether the true substrate has the same relationship with the enzyme as it does in the intact cell. I was wondering if there is any way of labelling the intact erythrocytes. Can you get ^{32}P into erythrocytes and see if the pattern is the same or not?

O.M. ROSEN: We tried doing that with $AT^{32}P$ and were not able to incorporate any ^{32}P into the isolated membranes.

G.H. DIXON: Did you try it with just inorganic ^{32}P phosphate?

O.M. ROSEN: No, we did not try just ^{32}P. We used $AT^{32}P$ with fluoride and glucose and a variety of other things, maybe ^{32}P would have been more effective. I presume that some of that $AT^{32}P$ ended up as ^{32}P anyhow, but we have not done that.

T.A. LANGAN: What is the effect of cyclic AMP on the auto phosphorylation of the regulatory subunit?

O.M. ROSEN: We were unable to show a significant effect of cyclic AMP on this process. It is conceivable that if we were able to slow down the process of phosphorylation significantly, we might see an effect, but we get a yield almost too quickly to look at a rate.

T.A. LANGAN: Do you mean that cyclic AMP does not block the phosphorylation, but you were unable to phosphorylate added regulatory subunit?

O.M. ROSEN: No, you can phosphorylate added regulatory subunits. You can also demonstrate phosphate incorporation

into the regulatory component of the whole cyclic AMP-dependent protein kinase in the absence of added cyclic AMP.
The studies of autophosphorylation have been performed just
in the last few weeks. One of the things that we plan to do
is to obtain phosphorylated binding protein, add it to the
catalytic subunit and see whether there is any difference
in the ability of the enzyme to reassociate, dissociate, and
so forth. These studies simply have not been done yet.

V. TOMASI: Does cyclic AMP have any effect on sodium-
potassium ATPase of erythrocyte membrane? And if so, do
you have any evidence relating phosphorylation of a fraction to sodium-potassium ATPase?

O.M. ROSEN: Yes, there have been studies showing that
protein III may be associated with an ATPase. This is not
our work, but is in the literature. If that is the case,
one might anticipate that you would get an acyl phosphate
intermediate in the position of protein III. We looked
specifically for this, and the phosphate incorporation
that we have observed does not appear to be in acyl linkage. We have not studied the effect of cyclic AMP on
ATPase activity.

H. SHEPPARD: I wonder if you could comment on the big
difference between the association constants for cyclic
AMP binding and activation of the protein kinase.

O.M. ROSEN: One possibility is that they were done at
different temperatures. The experiments to determine the
K_D were performed at 0° C and the activation constants
were derived from experiments performed at 35° C. We have
not looked at the effect of temperature systematically.

E.G. KREBS: How secure do you feel regarding the subunit
composition of the protein kinase?

O.M. ROSEN: I think at the moment it fits best with 2C,
2R (the 2R being one dimer of binding protein). I am not
sanquine, however, about the fact that we get apparently
only one mole of cyclic AMP bound per binding protein
dimer.

G.N. GILL: Would you repeat for us the stoichiometry of

the binding of cyclic AMP to the complex and to the isolated receptor proteins?

O.M. ROSEN: We find no difference between the cyclic AMP binding to the complex and to the isolated binding protein, and we get essentially one mole of cyclic AMP bound per 100,000 grams of binding protein.

G.N. GILL: So that would be 1 per dimer then?

O.M. ROSEN: Yes, 1 per dimer.

B.L. HORECKER: Have you tried to reconstitute a cAMP dependent protein kinase by mixing the isolated binding protein from the beef heart with the catalytic protein from erythrocytes?

O.M. ROSEN: No.

REGULATION OF THE MAMMALIAN PYRUVATE DEHYDROGENASE COMPLEX BY PHOSPHORYLATION AND DEPHOSPHORYLATION

L.J. REED, F.H. PETTIT, T.E. ROCHE, AND P.J. BUTTERWORTH
Clayton Foundation Biochemical Institute and Department
of Chemistry, The University of Texas at Austin

INTRODUCTION

Pyruvate dehydrogenase systems have been isolated from
Escherichia coli and from mitochondria of animal cells,
Neurospora crassa and Saccharomyces cerevisiae as functional
units with molecular weights in the millions (1). These
pyruvate dehydrogenase complexes have been separated into
three enzymes - pyruvate dehydrogenase, dihydrolipoyl
transacetylase, and dihydrolipoyl dehydrogenase, which act
in a coordinated manner as indicated in Fig. 1 (1,2).

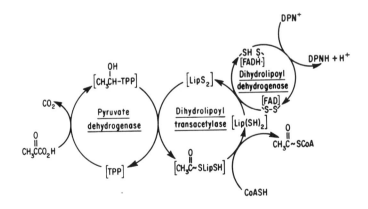

Fig. 1.   Reaction sequence in pyruvate oxidation.

Each of these complexes contains a core, consisting of dihydrolipoyl transacetylase, to which pyruvate dehydrogenase and dihydrolipoyl dehydrogenase are joined. Thus the transacetylase plays both a catalytic and a structural role. The mammalian complex also contains two regulatory enzymes, a kinase and a phosphatase, which are attached to the transacetylase (3). The mammalian dihydrolipoyl transacetylase as seen in the electron microscope has the appearance of a pentagonal dodecahedron, and its design appears to be based on icosahedral (532) symmetry (Fig. 2).

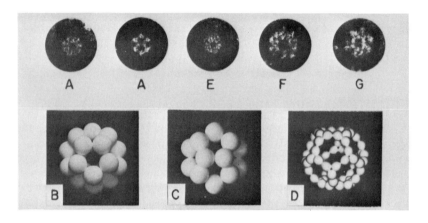

Fig. 2. Electron micrograph images and interpretative model of the mammalian dihydrolipoyl transacetylase. (A) Individual images (X250,000) showing two orientations of the bovine kidney transacetylase. (B,C) Corresponding views of a pentagonal dodecahedron photographed down a 5-fold axis and a 2-fold axis, respectively. Each sphere represents a group of three polypeptide chains. (D) Expanded model of the transacetylase showing trimer clustering at the twenty vertices. (E) Image (X250,000) of the bovine heart transacetylase viewed down a 5-fold axis. (F, G) Individual images (X250,000) of the pyruvate dehydrogenase complex from bovine kidney and heart, respectively. The electron micrographs were taken by Robert M. Oliver. The samples were negatively stained with phosphotungstate.

The transacetylase consists of 60 apparently identical
polypeptide chains of molecular weight about 52,000 (4).
Each chain apparently contains one molecule of covalently
bound lipoic acid. The pyruvate dehydrogenase component of
the mammalian complex has a molecular weight of about
154,000 and possesses the subunit composition $\alpha_2\beta_2$. Dihydrolipoyl dehydrogenase has a molecular weight of about
110,000 and contains two apparently identical polypeptide
chains and two molecules of FAD. The molecular weights of
the kinase and the phosphatase are about 50,000 and 100,000,
respectively. The subunit composition of the bovine kidney
and heart pyruvate dehydrogenase complexes is 60 transacetylase chains, 20-30 pyruvate dehydrogenase tetramers,
about 12 flavoprotein chains, about 5 kinase chains, and
about 5 phosphatase chains. Clearly, the complex does not
have its different components present in equimolar ratios.

RESULTS AND DISCUSSION

The activity of the purified mammalian pyruvate dehydrogenase complex is inhibited by the products of pyruvate oxidation, acetyl-CoA and DPNH, and these inhibitions
are reversed by CoA and DPN, respectively (5-7). These
observations have led to speculation that acetyl-CoA/CoA
and DPNH/DPN ratios may regulate the activity of the mammalian pyruvate dehydrogenase complex in vivo. However,
definitive evidence in support of this possibility remains
to be obtained.

Another regulatory mechanism, involving phosphorylation
and dephosphorylation of the pyruvate dehydrogenase component of the mammalian pyruvate dehydrogenase complex
(Fig. 3), was uncovered in this laboratory (8,9). We found
that highly purified preparations of the bovine kidney
pyruvate dehydrogenase complex were inactivated by incubation with micromolar concentrations of ATP. AMP, ADP, CTP,
GTP, and UTP were ineffective. Using radioactive ATP
labeled with ^{32}P in either the α-, $\alpha\beta$-, or γ-phosphoryl
moieties, we established that the terminal phosphoryl
moiety of ATP is transferred to the pyruvate dehydrogenase
complex. When the phosphorylated (inactivated) complex was
resolved, essentially all of the protein-bound radioactivity
was found in the pyruvate dehydrogenase component. Further
investigation revealed that the phosphorylated, inactivated

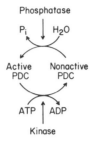

Fig. 3. Interconversion of active
and nonactive (phosphorylated) forms of
the mammalian pyruvate dehydrogenase
complex (PDC).

pyruvate dehydrogenase complex was reactivated by incubation
with millimolar concentrations of Mg^{2+}. Restoration of
activity was accompanied by release of inorganic ortho-
phosphate. Subsequent studies provided evidence that phosphorylation and concomitant inactivation of pyruvate dehydrogenase are catalyzed by a kinase (i.e., pyruvate dehydrogenase kinase) and dephosphorylation and concomitant reactivation are catalyzed by a phosphatase (i.e., pyruvate
dehydrogenase phosphatase). The data presented in Fig. 4
illustrate the time course of reciprocal changes in enzymic
activity and protein-bound phosphoryl groups obtained with
preparations of the pyruvate dehydrogenase complex from
mitochondria of bovine heart, kidney, and brain, and
porcine liver. The differences in rates of inactivation
and reactivation are apparently due to differences in the
amounts and possibly the activities of the kinase and the
phosphatase. Similar observations have been made with
preparations of the pyruvate dehydrogenase complex from
porcine heart and brain (10,11) and from rat epididymal
adipose tissue (12,13).

The pyruvate dehydrogenase component of the bovine
kidney and heart pyruvate dehydrogenase complex possesses
the subunit composition $\alpha_2\beta_2$. The molecular weights of the
two polypeptide chains are about 41,000 and 36,000 respectively (4). We have shown that the α-subunit, but not the
β-subunit, undergoes phosphorylation when the pyruvate dehydrogenase complex or the uncomplexed pyruvate dehydro-

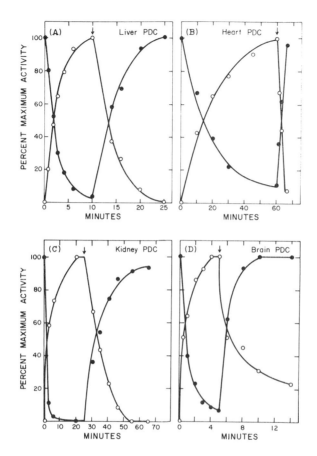

Fig. 4. Time course of phosphorylation and dephosphorylation of purified pyruvate dehydrogenase
complexes (PDC). The reaction mixtures contained
20 mM phosphate buffer, pH 7.0-7.5, 0.5 or 1.0 mM
$MgCl_2$, 2 mM dithiothreitol, 0.01-0.03 mM (A,B,C) or
0.5 mM (D) $[\gamma\text{-}^{32}P]ATP$, and enzyme complex. The mixtures were incubated at 25° (A,C) or 30° (B,D), and
aliquots were removed at the indicated times and
assayed for DPN-reduction activity (●) and for protein-bound radioactivity (o). At the time interval
indicated by the vertical arrow, sufficient $MgCl_2$
was added to give a final concentration of 10 mM
(A,B,C) or 20 mM (D).

genase is incubated with pyruvate dehydrogenase kinase and ATP. A radioactive tetradecapeptide has been isolated from tryptic digests of the ^{32}P-labeled pyruvate dehydrogenase, and its amino acid sequence has been determined (14).

Tyr-His-Gly-His-Ser(P)-Met-Ser-Asn-Pro-Gly-Val-Ser(P)-Tyr-Arg

The phosphoryl moieties are attached to seryl residues. The first seryl residue in this sequence is rapidly phosphorylated, and this phosphorylation results in inactivation of the pyruvate dehydrogenase complex. The third seryl residue is slowly phosphorylated. The physiological significance, if any, of this latter phosphorylation site remains to be determined.

Pyruvate dehydrogenase catalyzes both the decarboxylation of pyruvate to produce α-hydroxyethylthiamine-PP (Reaction 1) and the reductive acetylation of the lipoyl moieties, which are covalently bound to the transacetylase (Reaction 2). Phosphorylation of pyruvate dehydrogenase

$$CH_3COCO_2H + thiamine\text{-}PP \rightarrow CH_3CHOH\text{-}thiamine\text{-}PP + CO_2 \quad (1)$$

$$CH_3CHOH\text{-}thiamine\text{-}PP + [lipS_2] \rightarrow$$
$$[CH_3CO\text{-}S\text{-}lipSH] + thiamine\text{-}PP \quad (2)$$

with the kinase and ATP markedly inhibits the first reaction, but does not inhibit the second reaction (15). Since the α-subunit, but not the β-subunit of pyruvate dehydrogenase undergoes phosphorylation, these results suggest that the α-subunit catalyzes Reaction 1 and the β-subunit catalyzes Reaction 2. Thiamine-PP reduces the rate of phosphorylation of the pyruvate dehydrogenase complex by the kinase and ATP. Phosphorylation of the complex increases the K_D of the pyruvate dehydrogenase-Mg-thiamine-PP complex about twelvefold. It appears that the thiamine-PP binding site and the phosphorylation site on pyruvate dehydrogenase influence each other and that α-hydroxyethylthiamine-PP is bound to pyruvate dehydrogenase in a different orientation or possibly at a different site than is thiamine-PP.

It is interesting to note that the pyruvate dehydrogenase component of the E. coli pyruvate dehydrogenase complex, in contrast to the mammalian pyruvate dehydrogenase, consists of two apparently identical polypeptide chains of molecular weight about 96,000 (16,17). Moreover, the E. coli pyruvate dehydrogenase is not subject to regulation by phosphorylation and dephosphorylation (18). It appears that the E. coli pyruvate dehydrogenase has the two catalytic sites corresponding to Reactions 1 and 2 in a large polypeptide chain, i.e., a bifunctional chain, whereas the mammalian version of pyruvate dehydrogenase has the two catalytic sites segregated on two smaller, nonidentical chains. Recent results (C. R. Barrera and L. J. Reed, unpublished observations) indicate that the pyruvate dehydrogenase component of the pyruvate dehydrogenase complex isolated from mitochondria of S. cerevisiae resembles the mammalian rather than the E. coli pyruvate dehydrogenase in that it contains two nonidentical subunits. Although we have not been able to detect pyruvate dehydrogenase kinase or phosphatase activities in S. cerevisiae, the pyruvate dehydrogenase complex from S. cerevisiae is inactivated by incubation with bovine kidney pyruvate dehydrogenase kinase and ATP, and the phosphorylated complex is dephosphorylated, with concomitant reactivation, by bovine heart pyruvate dehydrogenase phosphatase and Mg^{2+}.

Various data indicate that the mitochondrial pyruvate dehydrogenase kinase and phosphatase are distinctly different from the cytosolic protein kinases and phosphatases. Thus the bovine kidney pyruvate dehydrogenase kinase does not catalyze a phosphorylation of histone, in the presence or absence of cyclic AMP (T. E. Roche and L. J. Reed, unpublished observations). Furthermore, a sample of skeletal muscle cyclic AMP-dependent protein kinase, kindly furnished by Dr. Edwin Krebs, did not inactivate preparations of the bovine kidney or heart pyruvate dehydrogenase complexes in the presence of ATP and cyclic AMP. The cyclic AMP-dependent protein kinase exhibited little ability, if any, to phosphorylate either the bovine kidney or heart pyruvate dehydrogenase complex or the crystalline pyruvate dehydrogenase from bovine kidney. The purified pyruvate dehydrogenase phosphatase showed little activity, if any, toward p-nitrophenyl phosphate. In our investigations, no effect of cyclic AMP was observed on the activities of the pyruvate dehydrogenase kinase or the phosphatase, even though preparations of the two enzymes were tested at various stages of purification, including crude mitochondrial extracts (19). Wieland and Siess (10) reported a cyclic AMP stimulation of the activity of the porcine heart pyruvate dehydrogenase phosphatase, which they attributed to a hypothetical cyclic AMP-dependent pyruvate dehydrogenase phosphatase kinase. Since there has been no confirmation of these results, in either our studies or those of other investigators (12,13), and particularly since Siess and Wieland (20) have qualified the previous interpretation of their observations, a direct role of cyclic AMP in regulation of the activity of pyruvate dehydrogenase phosphatase seems unlikely.

Some of the kinetic and regulatory properties of pyruvate dehydrogenase kinase and pyruvate dehydrogenase phosphatase have been determined. The true substrate for pyruvate dehydrogenase kinase is $MgATP^{2-}$, and the apparent K_m of the bovine kidney and heart kinases is about 0.02 mM (19). ADP is competitive with ATP, and the apparent K_i value for ADP is about 0.1 mM. Magnesium ion is required for pyruvate dehydrogenase phosphatase activity. The apparent K_m for Mg^{2+} is about 2 mM. The phosphatase is also active with Mn^{2+} (apparent K_m about 0.5 mM). Recent studies indicate that Ca^{2+} is required in addition to Mg^{2+} for pyruvate dehydrogenase phosphatase activity. Thus phosphatase activity is inhibited by EGTA in the presence of Mg^{2+} (19-21). Randle and coworkers (21) have shown that the Mg^{2+}-requiring pyruvate dehydrogenase phosphatase from pig heart, pig kidney cortex, and rat fat-cell mitochondria is markedly stimulated by Ca^{2+} at physiological concentrations (0.1-10 μM). We have found that Ca^{2+} is required to bind the phosphatase, but not the kinase or pyruvate dehydrogenase, to the transacetylase, thereby facilitating the Mg^{2+}-dependent dephosphorylation of the phosphorylated pyruvate dehydrogenase (22). The activity of the pyruvate dehydrogenase phosphatase from bovine kidney and heart is increased about tenfold when it is attached to the transacetylase. Ca^{2+} lowers the apparent K_m of the phosphatase for phosphorylated PDH about twentyfold (from about 58 μM to about 2.9 μM). The kinase is tightly bound to the transacetylase. We have separated these two enzymes and have shown that the transacetylase markedly stimulates the rate of phosphorylation of pyruvate dehydrogenase by the kinase

(19). The transacetylase lowers the apparent K_m of the kinase for pyruvate dehydrogenase from about 20 µM to about 0.6 µM. It appears that specific binding of the kinase, the phosphatase, and pyruvate dehydrogenase to the transacetylase is necessary to elevate the efficiency of the phosphorylation-dephosphorylation cycle to a physiologically useful level. Even so, the molecular activities of the bound kinase and phosphatase are relatively low, i.e., about 5 moles/min/mole of enzyme (19,20).

Pyruvate protects the pyruvate dehydrogenase complex against inactivation by ATP, and this effect appears to be more pronounced with the bovine heart complex than with the bovine kidney complex (9,19). The apparent K_i values for pyruvate are about 0.08 mM and 0.9 mM, respectively. Pyruvate is noncompetitive with ATP. We have obtained evidence that pyruvate exerts its inhibitory effect directly on the kinase (19).

Studies with the purified pyruvate dehydrogenase system suggest that the activity of pyruvate dehydrogenase kinase may be regulated in vivo by the intramitochondrial concentration of pyruvate and the ATP/ADP ratio and that the activity of the phosphatase may be regulated by the intramitochondrial concentrations of uncomplexed Mg^{2+} and Ca^{2+}. The concentration of uncomplexed Mg^{2+} and Ca^{2+} in the mitochondrial matrix may be determined, at least in part, by the ATP/ADP ratio, since ADP forms a much weaker complex with these divalent cations than does ATP. The Ca^{2+}-controlled association of the phosphatase and the transacetylase could provide an important mechanism for regulation of the phosphorylation-dephosphorylation cycle.

The activity of the pyruvate dehydrogenase complex is probably not regulated in vivo in an "on"-"off" manner by the kinase and the phosphatase. Rather, we visualize that these two antagonistic regulatory enzymes attain a steady state of activity, which is dependent on the various factors mentioned above as well as factors yet to be determined. Fig. 5 shows the effects of varying the concentrations of Mg^{2+} and Ca^{2+} on the near steady state activity of the bovine kidney pyruvate dehydrogenase complex reached in the presence of the kinase, the phosphatase, and ATP. The marked difference in the activities suggests that

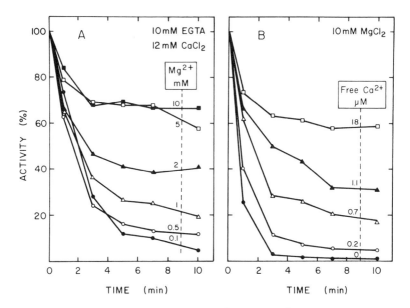

Fig. 5. Effects of Mg^{2+} and Ca^{2+} concentrations on the near steady state activity of the
bovine kidney pyruvate dehydrogenase complex
reached in the presence of the kinase, the phosphatase, and ATP. (A) The reaction mixtures contained 0.05 M 2-(N-morpholino)propanesulfonate
buffer, pH 7.4, 11 mM phosphate buffer, pH 7.4,
5 mM dithiothreitol, 0.2 mM ATP, 10 mM EGTA, 12 mM
$CaCl_2$, 0.44 mg of enzyme complex, 0.023 mg of
purified pyruvate dehydrogenase phosphatase, and
varying amounts of $MgCl_2$ in a total volume of
0.5 ml. (B) Components were as in (A) except
that 7.5 mM phosphate buffer and 10 mM $MgCl_2$
were present, and the concentration of $CaCl_2$ was
varied. The total concentration of $CaCl_2$ was 0,
7.4, 9.0, 9.4, and 10 mM. The concentrations of
free Ca^{2+} shown in the figure were calculated
from the association constant for CaEGTA (23).
ATP was added last, after the other components
had been incubated at 30° for 5 min. Aliquots
(0.02 ml) were removed at the indicated times
and assayed for DPN-reduction activity (19).

should the concentrations of free Mg^{2+} and Ca^{2+} in the mitochondrial matrix actually change during various metabolic states, the control exerted by these cations on pyruvate dehydrogenase phosphatase could be physiologically significant in regulating pyruvate dehydrogenase activity.

Evidence has been obtained which suggests that the phosphorylation-dephosphorylation cycle demonstrated with the purified pyruvate dehydrogenase system does indeed operate in intact mitochondria (24-26). Randle and co-workers (24) have reported that isolated rat epididymal fat-cell mitochondria showed an inverse relationship between ATP content and activity of the pyruvate dehydrogenase complex. These observations are consistent with competitive inhibition of pyruvate dehydrogenase kinase by ADP or possibly activation of pyruvate dehydrogenase phosphatase by Mg^{2+} and Ca^{2+}. At constant ATP concentration, pyruvate rapidly activated pyruvate dehydrogenase in rat fat-cell mitochondria, an observation consistent with inhibition of the pyruvate dehydrogenase kinase by pyruvate.

It appears that interconversion of the phosphorylated (nonactive) and nonphosphorylated (active) forms of pyruvate dehydrogenase is under hormonal and metabolic control. Insulin treatment of rat adipose cells increases the nonphosphorylated form, and this effect is antagonized by adrenaline and by adrenocorticotrophic hormone (12,13, 24). Metabolic states associated with increased concentrations of plasma free fatty acids result in an increase in the phosphorylated form of pyruvate dehydrogenase (27, 28). It would appear that these hormonal and metabolic effects on the interconversion of the phosphorylated and nonphosphorylated forms of pyruvate dehydrogenase are indirect. It seems possible that some of these effects may be mediated through changes in the intramitochondrial concentration of Ca^{2+}.

REFERENCES

(1) L.J. Reed, in: Current Topics in Cellular Regulation, Vol. 1, eds. B.L. Horecker and E.R. Stadtman (Academic Press, New York, 1969) p. 233.

(2) I.C. Gunsalus, in: The Mechanism of Enzyme Action,

eds. W.B. McElroy and B. Glass (Johns Hopkins Press, Baltimore, 1954) p. 545.

(3)   T.C. Linn, J.W. Pelley, F.H. Pettit, F. Hucho, D.D. Randall, and L.J. Reed, Arch. Biochem. Biophys. 148 (1972) 327.

(4)   C.R. Barrera, G. Namihira, L. Hamilton, P. Munk, M.H. Eley, T.C. Linn, and L.J. Reed, Arch. Biochem. Biophys. 148 (1972) 343.

(5)   P.B. Garland and P.J. Randle, Biochem. J. 91 (1964) 6c.

(6)   J. Bremer, Eur. J. Biochem. 8 (1969) 535.

(7)   O. Wieland, B. von Jagow-Westermann, and B. Stukowski, Hoppe-Seyler's Z. Physiol. Chem. 350 (1969) 329.

(8)   T.C. Linn, F.H. Pettit, and L.J. Reed, Proc. Nat. Acad. Sci. U.S.A. 62 (1969) 234.

(9)   T.C. Linn, F.H. Pettit, F. Hucho, and L.J. Reed, Proc. Nat. Acad. Sci. U.S.A. 64 (1969) 227.

(10) O. Wieland and E. Siess, Proc. Nat. Acad. Sci. U.S.A. 65 (1970) 947.

(11) E. Siess, J. Wittmann, and O. Wieland, Hoppe-Seyler's Z. Physiol. Chem. 352 (1971) 447.

(12) R.L. Jungas, Metabolism 20 (1971) 43.

(13) H.G. Coore, R.M. Denton, B.R. Martin, and P.J. Randle, Biochem. J. 125 (1971) 115.

(14) E.T. Hutcheson, Ph.D. Dissertation, University of Texas at Austin (1971).

(15) T.E. Roche and L.J. Reed, Biochem. Biophys. Res. Commun. 48 (1972) 840.

(16) O. Vogel and U. Henning, Eur. J. Biochem. 18 (1971) 103.

(17) M.H. Eley, G. Namihira, L. Hamilton, P. Munk, and L.J. Reed, Arch. Biochem. Biophys. 152 (1972) 655.

(18) E.R. Schwartz and L.J. Reed, Biochemistry 9 (1970) 1434.

(19) F. Hucho, D.D. Randall, T.E. Roche, M.W. Burgett, J.W. Pelley, and L.J. Reed, Arch. Biochem. Biophys. 151 (1972) 328.

(20) E.A. Siess and O.H. Wieland, Eur. J. Biochem. 26 (1972) 96.

(21) R.M. Denton, P.J. Randle, and B.R. Martin, Biochem. J. 128 (1972) 161.

(22) F.H. Pettit, T.E. Roche, and L.J. Reed, Biochem. Biophys. Res. Commun. 49 (1972) 563.

(23) H. Portzehl, P.C. Caldwell, and J.C. Rüegg, Biochim. Biophys. Acta 79 (1964) 581.

(24) B.R. Martin, R.M. Denton, H.T. Pask, and P.J. Randle, Biochem. J. 129 (1972) 763.

(25) S.M. Schuster and M.S. Olson, J. Biol. Chem. 247 (1972) 5088.

(26) S.M. Schuster and M.S. Olson, Biochemistry 11 (1972) 4166.

(27) O. Wieland, E. Siess, F.H. Schulze-Wethmar, H.G. von Funcke, and B. Winton, Arch. Biochem. Biophys. 143 (1971) 593.

(28) O.H. Wieland, C. Patzelt, and G. Löffler, Eur. J. Biochem. 26 (1972) 426.

This investigation was supported in part by Grant GM06590 from the U. S. Public Health Service.

DISCUSSION

J. ASHMORE: Do you think there is any basic difference in the pyruvate dehydrogenase complex between kidney, heart, liver and adipose tissue?

L.J. REED: You may recall that one of my slides (fig. 4) showed that purified preparations of the pyruvate dehydrogenase complex from kidney, heart, liver and brain exhibited the phosphorylation-dephosphorylation mechanism. Jungas, Randle, and Wieland and their co-workers have shown that this mechanism is operative with the pyruvate dehydrogenase system from rat epididymal adipose tissue. We have shown that the component enzymes of the bovine kidney and heart pyruvate dehydrogenase complexes are very similar. This is not to say that there are no differences in the properties of these two complexes. For example, the inhibitory effect of pyruvate on the kinase appears to be more pronounced with the heart complex than with the kidney complex.

H. SEGAL: I would like to come back to the question of the role of free fatty acids on the control of pyruvate dehydrogenase. As you pointed out elevated serum free fatty acids are correlated with low pyruvate dehydrogenase and vice versa. This suggests a relationship between hormone sensitive lipase and pyruvate dehydrogenase activities, and it invokes all of the regulatory mechanisms of the former as possible mediators, albeit indirectly, of the latter. You indicated that you thought that the role of free fatty acids in regulating the PDH system was indirect. Does that mean that you have been unable to find an effect of fatty acids either as such or as their CoA derivatives as activators of the kinase or as inhibitors of the phosphatase?

L.J. REED: In our hands, palmityl CoA and acetyl CoA have shown little effect, if any, on either the pyruvate dehydrogenase kinase or the pyruvate dehydrogenase phosphatase. However, acetyl CoA does inhibit the overall oxidation of pyruvate by the pyruvate dehydrogenase complex. A recent paper by Martin et al. (Biochem. J. 129 (1972) 763) reports that prostaglandin E_1, 5-methylpyrazole-3-carboxylate, and nicotinate, which are as effective as insulin as inhibitors of lipolysis, did not activate pyruvate dehydrogenase in epididymal fat pads.

96

We have not investigated pyruvate dehydrogenase activity in intact mitochondria or in intact cells. Therefore, my remarks must be limited to what has been reported by other investigators. It is my impression that the intimate connection between levels of plasma free fatty acids and between insulin and the activity of the pyruvate dehydrogenase complex remains to be determined.

E. REIMANN: I wondered if phosphorylase kinase acts as pyruvate dehydrogenase kinase as well?

L.J. REED: We have not tested phosphorylase kinase.

D. STEINBERG: It is possible that free fatty acids act through chelation of metal ions, particularly calcium and thereby exert an effect on the pyruvate dehydrogenase complex.

L.J. REED: That possibility does not appeal to me, but it is possible.

CYCLIC AMP-DEPENDENT PROTEIN KINASES FROM RABBIT RED BLOOD CELLS: ACTIVATION, FUNCTION, AND MOLECULAR FORMS

MARIANO TAO, RAJ KUMAR, AND PATRICIA HACKETT
Department of Biological Chemistry
University of Illinois at the Medical Center
Chicago, Illinois 60612

Abstract: Three forms of cyclic AMP-dependent protein kinases have been isolated from rabbit erythrocytes. By means of conventional enzyme purification procedures, kinase I has been enriched over four thousand-fold while kinases IIa and IIb have been purified nearly two thousand-fold, yielding functionally homogeneous proteins. They have comparable substrate specificities, preferably phosphorylating calf thymus histones. Their kinetic parameters are also remarkably similar with apparent Km values for cyclic AMP between 10^{-8} and 10^{-7} M, and for ATP and histone in the region of 10^{-5} M and 0.25 mg/ml, respectively. Kinase I is more labile to heat than kinases IIa and IIb. The molecular weights of kinases I, IIa, and IIb are estimated by gel-filtration to be 170,000, 120,000, and 240,000, respectively. Their corresponding sedimentation coefficients are 7.4, 5.2, and 7.2 S as determined by sucrose density gradient centrifugation. Cyclic AMP dissociates the kinases to their component catalytic and regulatory subunits which interact reversibly. The regulatory subunit of kinase I cross reacts with the catalytic moiety of both kinases IIa and IIb. Protamine also is capable of dissociating the kinases into their respective catalytic and regulatory components.

Two major protein fractions which exhibit phosphate acceptor activity have been isolated from rabbit erythrocytes and their specificities towards the kinases have been verified. The ability of the kinases to catalyze the phosphorylation of certain viral proteins has also been demonstrated.

99

INTRODUCTION

Cyclic AMP-dependent protein kinase has been established as one of the links in the sequence of reactions between
hormonal stimulation of adenylate cyclase and increased
glycogenolysis in rabbit skeletal muscle (1). In this system, the same kinase appears to catalyze the phosphorylation and activation of phosphorylase b kinase and the conversion of glycogen synthetase from the I form to the D
form (2-4). The occurrence of cyclic AMP-dependent protein
kinases seems to be ubiquitous (5). Furthermore, multiple
forms of these enzymes have been found in many of the
tissues investigated (6-9).

In addition to its involvement in glycogen metabolism,
cyclic AMP-dependent protein kinase also catalyzes the
phosphorylation of a wide variety of proteins (1,6,10-16).
This prompted Kuo and Greengard (5) to postulate that the
varied effects elicited by cyclic AMP are mediated through
the stimulation of protein kinases. However, this generalized concept appears to be premature since in Escherichia
coli the transcription of the genes of the lactose operon
is modulated by a cyclic AMP receptor protein devoid of
kinase activity (17).

The mechanism of activation of these kinases by the
cyclic nucleotide is a subject of considerable interest in
many laboratories including our own. We have previously
demonstrated (6) that cyclic AMP-dependent protein kinase
from rabbit red blood cells is an inactive complex of two
dissimilar functional subunits: a catalytic subunit and a
cyclic AMP-binding subunit (or regulatory subunit). The
activation of kinase is brought about by the binding of the
cyclic nucleotide to the regulatory subunit resulting in
its dissociation from the catalytic moiety as illustrated
in Fig. 1. The phosphotransferase activity is associated
only with the dissociated catalytic component. Since this
initial report, several other investigators have also
shown that cyclic AMP-dependent protein kinases isolated
from many tissues have similar molecular characteristics
(16-21). Based on the evidence accumulated to date, it is
tempting to propose that this mechanism of activation is
universal for all cyclic AMP-dependent protein kinases.
This novel mode of enzyme regulation deserves further in-

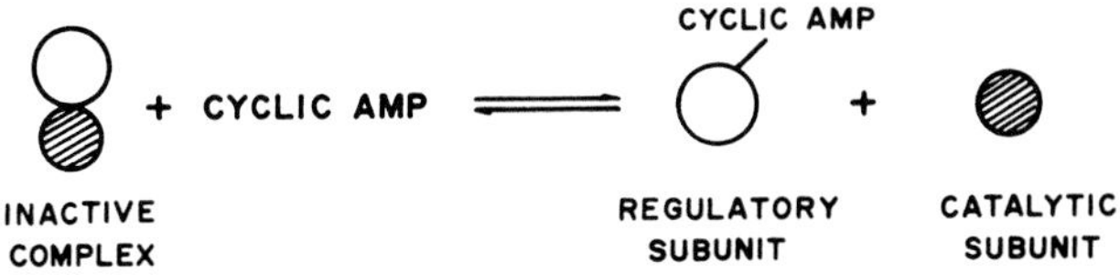

Fig. 1. Mode of action of cyclic AMP.

vestigation. We have therefore attempted to purify these
enzymes from rabbit red blood cells in order to study in
greater detail their properties as well as some of the factors which influence their subunit structure.

The chief function of the red cell is to transport
oxygen, and 95% of the dry weight of the cell is made up
of hemoglobin. The adult red cell is incapable of synthesizing macromolecules such as DNA, RNA, and proteins, due
to an absence of nucleus and other subcellular particles
(22). However, it still retains considerable cellular
complexity and an impressive array of enzymes, proteins,
lipids, etc., many of which participate in an active and
meaningful metabolic process. The presence of cyclic AMP-
dependent protein kinases in non-nucleated red cells is
intriguing and adds another dimension to the functional
properties of these cells. It also suggests that some of
the red cell metabolic processes may be under hormonal
control via the cyclic nucleotide. With this notion in
mind, we have also attempted to isolate and study the functional role of some of the red cell proteins which serve
as endogenous phosphate acceptors.

PURIFICATION OF CYCLIC AMP-DEPENDENT PROTEIN KINASES FROM
RABBIT RED BLOOD CELLS

Cyclic AMP-dependent protein kinases from rabbit red
blood cells are purified by conventional techniques
(Fig. 2). As shown in Fig. 3, chromatography on DEAE-cellulose resolves the enzyme activity into two major peaks,
I and II, with their corresponding cyclic AMP binding acti-

101

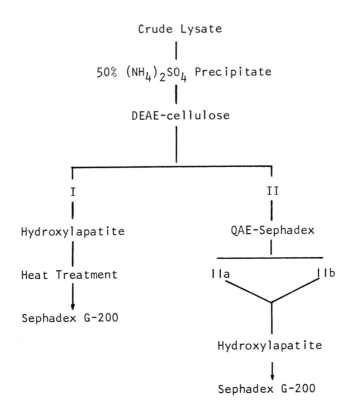

Fig. 2. Scheme of the purification procedure. The heat treatment is carried out at 53° C for 7 min in the presence of both ATP and Mg^{2+}.

vity. A similar elution profile from DEAE-cellulose chromatography is obtained with crude enzyme preparation from immature red blood cells (reticulocytes) as well (6). The kinase activity (or phosphotransferase activity) is measured both in the presence and absence of cyclic AMP by the incorporation of ^{32}P into calf thymus histone with $[\gamma-^{32}P]ATP$ as the phosphoryl donor (23). A convenient method for assaying the cyclic AMP binding activity is offered by Millipore filtration (24). Further fractionation of peak I consistently gives rise to a single component of cyclic AMP-dependent protein kinase activity. The functional homogeneity of this component is also revealed by

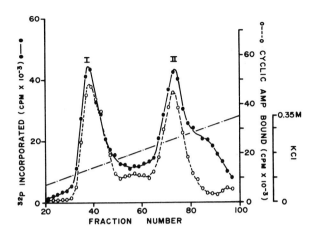

Fig. 3. Separation of rabbit erythrocyte cyclic AMP-dependent protein kinases on DEAE-cellulose column.

polyacrylamide gel electrophoresis. Peak I (hereon designated as cyclic AMP-dependent protein kinase I), isolated as outlined in Fig. 2, is enriched by about 4000-fold.

On the other hand, the peak II fraction from the DEAE-cellulose column may be further resolved by QAE-Sephadex chromatography into two cyclic AMP-dependent protein kinases, IIa and IIb (Fig. 4). A 2000-fold purification of both kinases IIa and IIb has been achieved. Analysis by polyacrylamide gel electrophoresis also reveals that both fractions are functionally homogeneous. Furthermore, in the case of IIa, evidence for a single protein band is obtained when stained with Coomassie blue.

PROPERTIES

An examination of the protein substrate specificity of red cell kinases shows a high degree of preference towards calf thymus histones as compared to albumin, casein, and protamine (TABLE 1). In addition, the crude fraction II from DEAE-cellulose column phosphorylates E. coli RNA polymerase (15). Our most recent experiment with kinase IIb indicates that this enzyme could also activate phosphorylase b kinase (25). TABLE 1 also shows that cyclic

AMP affects the phosphorylation reaction to varying degrees, depending upon the kinase and the protein substrate.

All three kinases are stimulated by low concentrations of cyclic AMP, and Km values between 10^{-8} and 10^{-7} M are obtained for the cyclic nucleotide. However, maximal stimulation could also be effected by cyclic nucleotides such as cyclic UMP, cyclic CMP, and cyclic GMP but at a much higher concentration ($Km \sim 10^{-5}$ M). Cyclic TMP, on the other hand, has no effect on any of the kinase reactions. The Km for ATP and histone for these kinases are unaffected by cyclic AMP and yield values in the region of 10^{-5} M and 0.25 mg/ml, respectively.

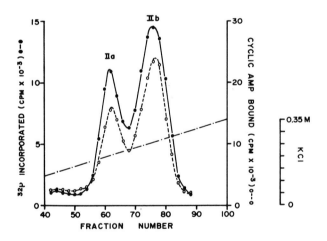

Fig. 4. Resolution of IIa and IIb by QAE-Sephadex chromatography.

TABLE 1

Phosphorylation of various proteins by red cell cyclic AMP-dependent protein kinases

Protein substrate	^{32}P Incorporated (pmoles)					
	Control			+ Cyclic AMP		
	I	IIa	IIb	I	IIa	IIb
Histone	84	99	237	687	182	445
Lysine-rich histone	23	64	153	818	275	582
Arginine-rich histone	11	29	48	182	64	141
Albumin	4	3	2	23	4	6
Casein	5	3	-	10	7	4
Protamine	12	2	16	19	7	17

The phosphorylation reactions are carried out as described previously (23) with a protein substrate concentration of 1.8 mg/ml.

TABLE 2

Heat inactivation of cyclic AMP-dependent protein kinases

Kinase	% Activity remaining after 10 min incubation at 53° C	
	In buffer alone	In buffer plus ATP and Mg^{2+}
I	11	89
IIa	50	32
IIb	65	54

Buffer: 0.01 M Tris-HCl, pH 7.5.

In an attempt to delineate possible differences between these kinases, we have examined the stability of these enzymes towards heat (53°) under various conditions as shown in TABLE 2. The phosphotransferase activity of both kinases IIa and IIb are inactivated at about the same rate when heated for 10 min at 53° C in 0.01 M Tris-HCl buffer, pH 7.5. Approximately 50% or less of the phosphotransferase activity is lost by kinases IIa and IIb under these conditions. In contrast, kinase I appears to undergo rapid inactivation at this temperature. However, the stability of kinase I is greatly enhanced by the presence of both ATP and Mg^{2+} in the heating mixture, whereas with IIa and IIb these compounds not only failed to afford any protective effect, but appear to slightly augment their rate of inactivation. The results indicate that although kinases IIa and IIb have approximately the same heat stability, they are distinct from kinase I.

QUATERNARY STRUCTURE

The molecular weights of cyclic AMP-dependent protein kinase I, IIa, and IIb have been estimated by Sephadex gel-

filtration to be 170,000, 120,000, and 240,000, respectively. When these kinases are examined by sucrose density
gradient centrifugation using ovalbumin (3.6 S) and E. coli
alkaline phosphatase (6.3 S) as standards, sedimentation
coefficients of 7.4, 5.2, and 7.2 S are obtained respectively for I, IIa, and IIb. The disparity between the sedimentation coefficient and the molecular weight obtained
from gel-filtration for IIb suggests that either the enzyme has unique physicochemical properties or that it may
have undergone dissociation. Limited proteolysis by contaminating enzymes may also account for this observation.
A similar anomalous behavior has been observed in the instance of bovine heart muscle cyclic AMP-dependent protein
kinase (21). Our preliminary observations further indicate
that upon storage in liquid nitrogen in the absence of re-
ducing agent, the enzyme IIb appears to be converted into
a 5.2 S sedimenting component, suggesting that kinase IIa
may have derived from IIb. However, definitive proof of
this must await further analysis.

Our earlier finding that cyclic AMP causes the dissociation of kinase I into regulatory and catalytic subunits
(6) prompted us to examine kinases IIa and IIb under similar conditions. The results presented in Fig. 5 and 6
show that an analogous behavior prevails with kinases IIa
and IIb. The cyclic nucleotide converts IIa (5.2 S) into
4.3 S catalytic and 3.4 S regulatory components, whereas
with IIb a sedimentation coefficient of 5.8 S is obtained
for both the catalytic and the binding activities. This
observation may be interpreted in terms of kinase IIb
being constructed from catalytic and regulatory subunits
of approximately equal sizes. Alternatively, this may represent an oligomer of lower order of the two functional
subunits. A similar observation has been reported by
Miyamoto and his associates (26) with bovine brain protein
kinase. These studies also point out that in these kinases, the regulatory subunit behaves as an inhibitory protein, which upon complex formation with cyclic AMP releases the catalytically active enzyme.

The interaction of the two dissimilar functional subunits appears to be reversible. Figure 7 shows that the
addition of varying amounts of the regulatory subunit of
kinase I to its catalytic moiety (CI) progressively inhi-

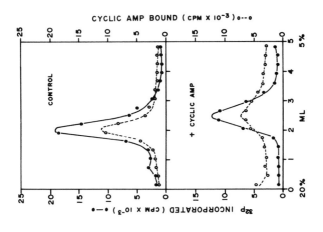

Fig. 6. Sedimentation of kinase IIb in the presence and absence of 10^{-6} M cyclic AMP.

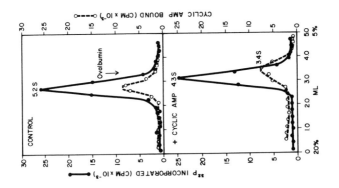

Fig. 5. Sedimentation of kinase IIa in the presence and absence of 10^{-6} M cyclic AMP.

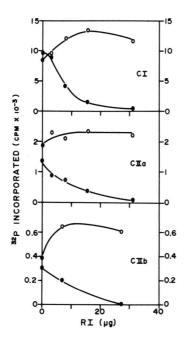

Fig. 7. Effect of the regulatory subunit of kinase I on the catalytic moiety of I, IIa, and IIb, in the presence (o——o) and absence (●——●) of cyclic AMP.

bits the phosphotransferase activity. Interestingly, the regulatory subunit of kinase I also cross-reacts with the catalytic moiety of both kinases IIa (CIIa) and IIb (CIIb). In all experiments, cyclic AMP prevents this inhibition by the regulatory component.

INTERACTION WITH PROTAMINE

Protamine, evidently, is a poor substrate for red cell cyclic AMP-dependent protein kinases; and its phosphorylation shows little, if any, stimulation by cyclic AMP (TABLE 1). The phosphorylation of this substrate by liver cyclic AMP-dependent protein kinase is also independent of the cyclic nucleotide (27). In an attempt to explain

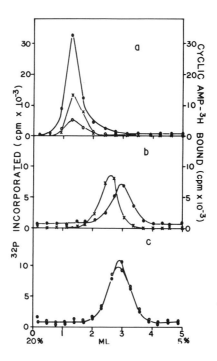

Fig. 8. Sucrose density gradient centrifugation of kinase I under various conditions (23): (a) control, (b) in presence of 10^{-6} M cyclic [3H] AMP, (c) preincubated with 0.6 mg/ml protamine. Kinase activity: 0——0 -cyclic AMP; ●——●, +cyclic AMP. Binding activity: X X.

this selective effect of cyclic AMP on the phosphorylation of different protein substrates, we have performed a more detailed analysis of the interaction between protamine and red cell kinases.

When an incubation mixture of protamine and kinase I is sedimented in a 5 - 20% sucrose density gradient, a single peak of cyclic AMP-independent kinase activity is

110

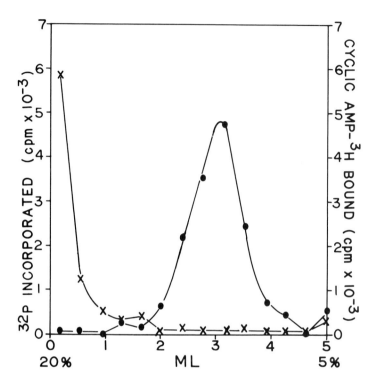

Fig. 9. Sedimentation of kinase I preincubated with protamine and cyclic [3H]AMP in a sucrose gradient containing 10^{-6} M cyclic [3H]AMP (23). ●——●, kinase activity; X——X, binding activity.

obtained as shown in Fig. 8c. This activity sediments at a position similar to the catalytic moiety obtained from the dissociation by cyclic AMP indicating that protamine also causes the dissociation and activation of kinase I (23). Unfortunately, the cyclic AMP binding activity in the same gradient (Fig. 8c) cannot be measured due to in-activation of this component. However, if a similar sedi-mentation experiment is carried out by including cyclic AMP in the gradient, a heavy sedimenting cyclic AMP bind-

111

ing component can be detected at the bottom of the tube
(Fig. 9). The presence of cyclic AMP in the gradient apparently enhances the stability of the regulatory subunit
(24). On the basis of these observations, the lack of
stimulation by cyclic AMP of the phosphorylation of protamine (or other protein substrates) may be attributed to
the ability of this protein substrate to complex with the
regulatory subunit and to release the catalytic activity.
In contrast to protamine, histone "mixture" and lysine-
rich histone are both unable to dissociate protein kinase
I.

An analogous effect of protamine on kinases IIa and
IIb has been observed. In both instances, protamine
causes the regulatory subunit of these kinases to sediment
to the bottom of the tube. The sedimentation coefficient
of the catalytic moiety of kinase IIa (4.1 S) obtained
from this experiment is similar to that found by centrifugation in a gradient containing cyclic AMP. However, on
incubation with protamine, the catalytic moiety of IIb
sediments at 4 S whereas in the presence of cyclic AMP
its sedimentation coefficient is 5.8 S (Fig. 10). A somewhat similar behavior has been reported with bovine brain
protein kinase by Miyamoto et al. (26). Several interpre-
tations are possible for these observations. Kinase IIb
may be assumed to have a unique quaternary structure as
discussed previously. The slower-sedimenting catalyic component detected with protamine, therefore, would represent
a complete separation of the two dissimilar functional
subunits. An alternative interpretation would be that the
5.8 S catalytic moiety is derived from the 4 S component
by aggregation with itself or with contaminating proteins.
However, this would suggest that protamine is capable of
interfering with this type of aggregation. Unfortunately,
because of the inherent limitations in determining molecular weights of proteins by sucrose density gradient
centrifugation, a meaningful evaluation of the data remains hazardous at this time.

ENDOGENOUS PHOSPHATE ACCEPTORS IN THE RABBIT RED BLOOD CELLS

Several laboratories have identified endogenous protein substrates for the corresponding tissue protein kina-

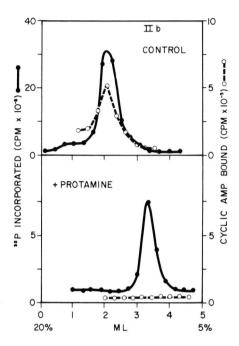

Fig. 10. Effect of preincubation with protamine on the sedimentation of IIb.

ses. Membrane proteins from ox brain (28), rat brain (13), and rat liver (29), bovine adrenal ribosomal proteins (14), microtubule of bovine brain (16), ribosomal proteins of rabbit reticulocytes (30) and rat liver (31), and most recently, the membrane proteins of human erythrocytes (32, 33) have all been shown to be phosphate acceptors. However, stimulation by cyclic AMP has been demonstrated only in some of these instances.

In our efforts to explore the biological function and significance of the kinases in the rabbit red blood cell, we have isolated and partially purified the soluble endogenous substrates for these enzymes. This should enable us to examine the behavior and properties of the kinases

under situations where they are furnished with their natural substrates instead of an added exogenous phosphate
acceptor like histone.

As shown in Fig. 11, two protein fractions containing
phosphate accepting activity are eluted from a DEAE-cellulose column with KCl gradient. Acceptor protein I (API)
is eluted at a position between the two kinase activity
fractions (I and II), while acceptor protein II (APII)
emerges from the column immediately after kinase fraction
II. Both of these acceptors are further purified by Sephadex G-200 gel-filtration, and their ability to accept ^{32}P
from $[\gamma\text{-}^{32}P]ATP$ in the presence of kinase is verified.
The effect of cyclic AMP on the phosphorylation of these
endogenous proteins is also investigated. The data summarized in TABLE 3 indicate that API contains a consider-

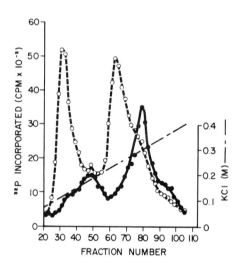

Fig. 11. Resolution of endogenous phosphate acceptors
and protein kinases on DEAE-cellulose column. 0----0,
protein kinase activity. ●——●, acceptor activity.

TABLE 3

Phosphorylation of endogenous proteins by rabbit red blood cell cyclic AMP-dependent protein kinases

Addition	^{32}P Incorporated (pmoles)	
	Control	+ Cyclic AMP
AP I	133	151
AP I + kinase I	122	158
AP I + kinase IIa	130	159
AP I + kinase IIb	118	175
AP II	36	45
AP II + kinase I	43	79
AP II + kinase IIa	45	73
AP II + kinase IIb	50	90

The phosphorylation reactions are carried out as described previously (23) except histone is replaced by either AP I (0.8 mg/ml) or AP II (0.7 mg/ml). The enzyme concentrations for kinases I, IIa, and IIb are 50, 100, and 55 µg/ml respectively.

able amount of kinase activity and that cyclic AMP has only a slight effect on the phosphorylation reaction. The substrate specificity of AP I with respect to the three kinases is also not discernible, possibly due to the high contamination of AP I with kinases. AP II, on the other hand, has less endogenous kinase activity; and cyclic AMP stimulates its phosphorylation. Furthermore, all three kinases are equally effective in phosphorylating AP II. A more detailed characterization of these protein acceptors is in progress.

115

PHOSPHORYLATION OF VIRAL PROTEINS

In collaboration with Dr. Walter Doerfler, we have demonstrated that the crude fraction II from the DEAE-cellulose column also phosphorylates the polypeptides of adeno-
virus types 2 and 12 and bacteriophage λ but not bacteriophage T4 and tobacco mosaic virus (34). Analysis by SDS-
polyacrylamide gel electrophoresis indicates that at least
three or four of type 2 adenoviral associated proteins are
phosphorylated. More recently, protein kinase activity has
been detected in purified virions (35, 36). In addition,
analysis of simian virus 40 structural proteins reveals
that all of them contain phosphates (37). The biological
function of the phosphorylation of viral proteins remains
to be established but may very well entail a mechanism to
regulate the configuration and function of these proteins
during biosynthesis and assembly of the virion.

CONCLUSIONS

At least three cyclic AMP-dependent protein kinases
separable by conventional enzyme fractionation techniques
are found in rabbit red blood cells. The quaternary structure of these enzymes appears to be similar, consisting of
regulatory and catalytic subunits. Cyclic AMP activates
these enzymes by binding to the inhibitory regulatory
moiety and concomitantly releasing the catalytically active
component. Interestingly, this process could also be
brought about by protamine which seems to have a strong
affinity for the regulatory component. The interrelation-
ship of these enzymes is yet to be resolved. Several laboratories (38, 39) have suggested that these multiple enzymic forms may in part arise from proteolytic action. In
our hands, such a possibility may account for the existence
of IIa. However, this is based on fragmentary evidence
and must await further analysis. On the basis of heat stability, kinase I appears to be distinct from IIa and IIb.

Of no less interest is the functional role of these
enzymes in red blood cells. These enzymes probably do not
play a significant role in the regulation of glycogen synthesis and degradation since the metabolism of glycogen in
these cells appears to be minimal (22). Endogenous phosphate acceptor proteins for these kinases are present in

red blood cells, but the metabolic role of these proteins has not yet been established. The importance of these kinases in red blood cells may be elucidated only after the delineation of the role of these acceptor proteins.

REFERENCES

(1) D.A. Walsh, J.P. Perkins and E.G. Krebs, J. Biol. Chem., 243 (1968) 3763.

(2) C. Villar-Palasi and K.K. Schlender, Fed. Proc., 29 (1970) 938.

(3) T.R. Soderling, J.P. Hickenbottom, E.M. Reimann, F.L. Hunkeler, D.A. Walsh and E.G. Krebs, J. Biol. Chem., 245 (1970) 6317.

(4) F. Huijing and J. Larner, Proc. Nat. Acad. Sci. U.S.A., 56 (1966) 647.

(5) J.F. Kuo and P. Greengard, ibid., 64 (1969) 1349.

(6) M. Tao, M.L. Salas and F. Lipmann, ibid., 67 (1970) 408.

(7) E.M. Reimann, D.A. Walsh and E.G. Krebs, J. Biol. Chem., 246 (1971) 1986.

(8) A. Kumon, K. Nishiyama, H. Yamamura and Y. Nishizuka, ibid., 247 (1972) 3726.

(9) F. Lipmann, Advan. Enzyme Regul., 9 (1971) 5.

(10) T.A. Langan, Science, 162 (1968) 579.

(11) J.D. Corbin and E.G. Krebs, Biochem. Biophys. Res. Commun., 36 (1969) 328.

(12) J.E. Loeb and C. Blat, FEBS Letter, 10 (1970) 105.

(13) E.M. Johnson, H. Maeno and P. Greengard, J. Biol. Chem., 246 (1971) 7731.

(14)   G.M. Walton, G.N. Gill, I.B. Abrass and L.D. Garren, Proc. Nat. Acad. Sci. U.S.A., 68 (1971) 880.

(15)   O.J. Martelo, S.L.C. Woo, E.M. Reimann and E.W. Davie, Biochemistry, 9 (1970) 4807.

(16)   D.B.P. Goodman, H. Rasmussen, F. DiBella and C.E. Guthrow, Jr., Proc. Nat. Acad. Sci. U.S.A., 67 (1970) 652.

(17)   R.L. Perlman and I. Pastan, Current Topics in Cellular Regulation, 3 (1971) 117.

(18)   G.N. Gill and L.D. Garren, Biochem. Biophys. Res. Commun., 39 (1970) 335.

(19)   A. Kumon, H. Yamamura and Y. Nishizuka, ibid., 41 (1970) 1290.

(20)   E.M. Reimann, C.O. Brostrom, J.D. Corbin, C.A. King and E.G. Krebs, ibid., 42 (1971) 187.

(21)   C.S. Rubin, J. Erlichman and O.M. Rosen, J. Biol. Chem., 247 (1972) 36.

(22)   J.W. Harris and R.W. Kellermeyer, The Red Cell (Harvard University Press, Boston, 1970).

(23)   M. Tao, Biochem. Biophys. Res. Commun., 46 (1972) 56.

(24)   M. Tao, Arch. Biochem. Biophys., 143 (1971) 151.

(25)   N. Ressler and M. Tao, unpublished observation.

(26)   E. Miyamoto, G.L. Petzold, J.S. Harris and P. Greengard, Biochem. Biophys. Res. Commun., 44 (1971) 305.

(27)   T.A. Langan, Ann. N.Y. Acad. Sci., 185 (1971) 166.

(28)   M. Weller and R. Rodnight, Nature, 225 (1970) 187.

(29)   L. Schlatz and G.V. Marinetti, Biochem. Biophys. Res. Commun., 45 (1971) 51.

(30)   D. Kabat, Biochemistry, 10 (1971) 197.

(31)   C. Eil and I. Wool, Biochem. Biophys. Res. Commun.,
       43 (1971) 1001.

(32)   C.S. Rubin, J. Erlichman and O.M. Rosen, J. Biol.
       Chem., 247 (1972) 6135.

(33)   C.E. Guthrow, Jr., J.E. Allen and H. Rasmussen, ibid.,
       247 (1972) 8145.

(34)   M. Tao and W. Doerfler, Eur. J. Biochem., 27 (1972)
       448.

(35)   M. Strand and J.T. August, Nature N. Biol., 233 (1971)
       137.

(36)   C.C. Randall, H.W. Rogers, D.N. Downer and G.A. Gentry,
       J. Virol., 9 (1972) 216.

(37)   K.B. Tan and F. Sokol, J. Virol., 10 (1972) 985.

(38)   A. Kumon, K. Nishiyama, H. Yamamura and Y. Nishizuka,
       J. Biol. Chem., 247 (1972) 3726.

(39)   J.D. Corbin, C.O. Brostrom, C.A. King and E.G. Krebs,
       ibid., 247 (1972) 7790.

       This research was supported in part by grants from the
American Cancer Society (BC-65), the National Science
Foundation (GB-27435A#1), and the Illinois Division of the
American Cancer Society (#73-1).

DISCUSSION

J. WILLIS: Are these kinases also present in reticulocytes?

M. TAO: Yes, at least if you process it through the DEAE-
cellulose column. You see a similar distribution profile
compared to that of the erthrocytes.

T. SODERLING: You showed one slide where increasing protamine concentration give you an increase in binding of cyclic AMP to the protein kinase. I wonder if you tried other proteins such as bovine serum albumin, because I believe they would do the same thing.

M. TAO: Yes, we have tried bovine serum albumin and we have found that it has no effect on the binding activity of protein kinases. We also have tried histones and they do not seem to be as effective as protamine.

T. SODERLING: You do your binding by the millipore technique?

M. TAO: Right, we want to confirm the effect of this protein on the binding activity using dialysis or sephadex gel-filtration. We have not done that yet.

T. SODERLING: In our binding assay that is essentially the same as Gilman's, we find bovine serum albumin is just as effective as the protein kinase inhibitor.

H. SHEPPARD: If you do not add cyclic AMP to the system, but simply add protamine and obtain a precipitate of the protamine-regulatory protein complex, can you now demonstrate cyclic AMP binding activity?

M. TAO: In the case of kinase I, it is difficult to demonstrate any binding activity after complex formation with protamine because of instability, but, we know that cyclic AMP has a protective effect on this subunit. With regards to IIa and IIb the binding activities seem to be a little bit more stable. These regulatory components sedimented to the bottom of the tube.

P. GREENGARD: Previous methods of preparing regulatory subunits of protein kinases have used cyclic AMP to dissociate the holoenzyme and have suffered from the disadvantage that it is extremely difficult to free the regulatory subunit from bound cyclic nucleotide. Recently, regulatory subunit free of bound cyclic AMP was prepared from brain protein kinase in our laboratory (Miyamoto et al., J. Biol. Chem., 248, 179; 1973) by taking advantage of the ability of histone (Miyamoto et al., Biochem. Biophys. Res. Comm., 44, 305; 1971) to cause the dissociation of the enzyme into sub-

units. For this purpose, a partially purified enzyme preparation from a DEAE-cellulose column was preincubated at
30° for 10 minutes in the presence of 100 µg of histone per
ml. After preincubation, the solution was applied to a
column of hydroxylapatite, the column washed, and the protein eluted. Cyclic AMP-binding activity was assayed on
aliquot of each fraction. Active fractions were combined
and used as cyclic AMP-binding protein. Protein kinase
catalytic activity of this preparation was negligible. Thus
dissociation of holoenzymes by protein substrates provides
an effective means of preparing regulatory subunits without
cyclic nucleotides being attached.

S. STRADA: Do the reticulocytes have the same membrane acceptor protein as the red blood cell?

M. TAO: I do not know. We have not tried reticulocytes.

H. SHEPPARD: I think it is important to make one comment.
While it is very useful to proceed with the search for a
function of the protein kinase or what ever else one finds
in red blood cells, I think it is very important to note
that these cells are decaying and are on their way to being
eliminated. Therefore these protein may represent residues
of physiological roles during the development and maturation
periods.

M. TAO: I agree with you.

E.G. KREBS: Continuing the discussion on the point that
was just made, do not most of the enzyme activities which
are very high in the reticulocyte drop precepitously, when
it becomes an adult red cell? What happens to the protein
kinase?

M. TAO: I am not sure to what extent the protein kinase
drops upon maturation of the reticulocyte but there is still
a quite significant kinase activity in the mature red cells.

PROTEIN KINASE IN THE BOVINE CORPUS LUTEUM

S. GOLDSTEIN and J. M. MARSH
Department of Biochemistry and The Endocrine Laboratory
University of Miami School of Medicine

Abstract: Cyclic AMP, endogenously synthesized in bovine
corpus luteum slices under stimulation of luteinizing hormone,
has been localized in the cytosol fraction of this tissue. A
portion of the cyclic AMP is bound to a macromolecule of
molecular weight greater than 30,000. Cyclic AMP dependent
protein kinase has also been localized in the cytosol and has
been partially purified from this fraction. Kinase activity
was assayed against several protein substrates during the
purification. The order of specific activities at any purification step was protamine sulfate > histone type II-A > casein;
however, the greatest purification and yield was achieved
with histone as substrate. The variability of purification and
yield of activity with these substrates is interpreted as indicating the presence of multiple protein kinases. Rabbit
skeletal muscle glycogen synthetase kinase activity was
assayed and a physical separation of this activity from the
other three kinase activities was achieved on a hydroxylapatite column. Cholesterol esterase activity was assayed in
slice preparations of corpora lutea and it was found that both
LH ($2\mu g/ml$) and cyclic AMP (0.02M) had slight but inconsistent stimulatory effects on cholesterol esterase activity
(p=0.1 and 0.1< p <0.2 respectively). The partially purified
protein kinase has been tested with cholesterol side-chain
cleavage preparations and cholesterol esterase preparations

but so far no consistent stimulatory effect on either of these systems has been found.

INTRODUCTION

Our laboratory has carried out an investigation over the past several years on the process of steroidogenesis in the corpus luteum and its control by luteinizing hormone (LH). The approach we used in our early work was to develop a model system in which slices of bovine corpora lutea were incubated in vitro and the effects of tropic hormones measured on several biochemical parameters.

We found that if slices of bovine corpora lutea were incubated in the presence of LH there was a marked increase in progesterone synthesis. This effect was specific for LH and quite sensitive to small amounts of the hormone, the minimum effective dose being approximately 10^{-10}M (1).

In our investigation of the mechanism of LH action, we found that cyclic AMP was probably a mediator of this gonadotropin's effect, in that LH increased the endogenous concentration of the cyclic nucleotide, and exogenous cyclic AMP mimicked the effect of the hormone in stimulating steroidogenesis (2, 3). We also found that the increase in cyclic AMP was due to an activation of the adenyl cyclase system rather than to an inhibition of the phosphodiesterase enzyme system (4).

SUBCELLULAR LOCALIZATION OF CYCLIC AMP

The site of action of LH and cyclic AMP in the steroidogenic pathway is believed to be somewhere between cholesterol and pregnenolone, but the exact mechanism by which cyclic AMP accelerates this conversion is unknown. In an attempt to shed some light on this mechanism, we began to investigate where was the endogenous cyclic AMP located in the cell and in what state did it exist.

Cyclic AMP was assessed using a modification of the method

of Kuo and deRenzo (5). Slices of a corpus luteum were incubated for 1 hour at 37°C, in Krebs-Ringer bicarbonate buffer, in the presence of 6 μCi of [8-^{3}H] adenine. The slices were then rinsed in isotonic NaCl and incubated again for 30 min under control conditions or in the presence of LH. At the end of the second incubation the slices were homogenized and the [8-^{3}H] cyclic AMP isolated by a modification of the procedure of Krishna et al. (6). The modification involved an additional chromatography of the ^{3}H labelled product in a cellulose thin layer system to achieve radiochemical purity. As is shown in Table 1 a significant portion of the material in the BaSO$_4$ precipitate is not [8-^{3}H] cyclic AMP, especially in samples prepared from control tissue. The material eluted from the thin layer system is nearly pure cyclic AMP as shown by similarity of the specific activities of cyclic AMP and 5'.-AMP in steps 2 and 3 respectively.

Using this technique we found that LH caused a marked increase in the accumulation of ^{3}H-cyclic AMP (Figure 1). This effect of LH on the accumulation of [8-^{3}H] cyclic AMP was statistically significant with a p value less than 0.001 and was very similar in magnitude to the effect we have observed previously on the accumulation of mass amounts of cyclic AMP. (2).

We then undertook an investigation to determine where the ^{3}H-cyclic AMP was localized in the cell. In 6 experiments, slices which had been incubated with [8-^{3}H] adenine and LH were homogenized in 0.25 M sucrose, containing 0.05 M Tris-HCl buffer, pH 7.4, 0.04 M theophylline, and 0.02 M EDTA. The homogenate was then separated by differential centrifugation using standard techniques into the five fractions. Most of the ^{3}H-cyclic AMP was localized in the 105,000 x g supernatant fraction and only small and equal amounts localized in the other fractions. (Figure 2) From the marker enzyme studies, particularly the glucose-6-phosphate dehydrogenase assay, it was apparent that this supernatant fraction contained primarily cytosol material.

It was considered that some of the cyclic AMP found in this fraction might have been removed from the particulate materials

125

TABLE 1

Test of the radiochemical purity of [8-^{3}H] cyclic AMP
isolated from corpora lutea tissue

STEP	Specific Activity	
	Control	LH
	dpm/μg	
1. Supernatant obtained after Dowex-50 chromatography and BaSO$_4$ precipitation	237	3300
2. Eluate from the cyclic AMP zone after 1st thin layer chromatography	67.3	3330
3. Eluate from the 5'-AMP zone after phosphodiesterase treatment and 2nd thin layer chromatography	50.2	3060

This table illustrates the changes in specific activity of
the [8-^{3}H] cyclic AMP isolated from a control incubation
of corpora lutea slices and an incubation with 2μg/ml of LH.
The ratio of dpm to the μg of carrier cyclic AMP was sequentially determined on the material in the supernatant of the
third precipitation with BaSO$_4$, on the material eluted from
the thin layer chromatogram which migrated with cyclic AMP,
and on the 5'-AMP material obtained after cyclic nucleotide
phosphodiesterase treatment and rechromatography on the
same thin layer system.

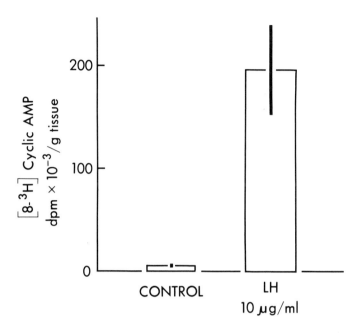

Fig. 1. The effect of luteinizing hormone on the accumulation of [8-³H] cyclic AMP from [8-³H]adenine in 10 paired incubations. The height of the bar represents the mean and the heavy line the standard error of the mean. Approximately 0.5 g of slices of a corpus luteum were incubated for 1 hour at 37° in an atmosphere of 95% O_2 and 5% CO_2 in 5 ml of Krebs-Ringer bicarbonate buffer containing 6μCi of 8-³H adenine. The slices were rinsed in 0.154M NaCl and incubated for 30 min at 37° under an atmosphere of 95% O_2 and 5% CO_2 in 5 ml of Krebs-Ringer bicarbonate buffer containing 0.01M theophylline or 5 ml of this buffer-theophylline solution plus 10 μg/ml of LH. At the end of the second incubation the slices were homogenized in 0.048M Tris-HCl, pH 7.4, and 0.01 M theophylline at 0°. The homogenate was heated at 100° for 2 min, carrier cyclic AMP (0.625 μmoles) was added, and the [8-³H] cyclic AMP isolated as described in the text.

of the cell by the repeated washing procedure that was used in the isolation of the particulate fractions. To assess this possibility, separate experiments were carried out in which the cytosol fraction was prepared immediately after homogenization by centrifuging the homogenate at 105,000 x g for 60 min and omitting the washing of the resultant pellet. The percentage of [8-^{3}H] cyclic AMP in the cytosol fraction in these experiments was essentially the same as in the former study, namely about 85% of the [8-^{3}H] cyclic AMP was in the cytosol fraction.

It was assumed that all the cyclic AMP measured in the particulate fractions was bound to those fractions in some way or present within the organelles, since it centrifuged down with the particules, even after repeated washings. In the cytosol fraction, however, it was not known if the [8-^{3}H] cyclic AMP was entirely in the free form or if any was bound to some other molecule. To assess this, the cytosol fractions from control and LH treated tissues were subjected to ultrafiltration in an Amicon cell, using a PM-30 membrane, which will retain a globular protein of approximately 30,000 molecular weight. After about 80% of the cell charge had passed through the membrane, the concentration of ^{3}H-cyclic AMP was determined in the filtrate and in the retentate and the amount of bound cyclic AMP calculated from these values. (Figure 3).

Almost all of the ^{3}H-cyclic AMP in the control cytosol appears to be bound to a macromolecule and about 25% of the labelled nucleotide is bound in the LH treated tissue. Although the percentage bound is greater in the control cytosol, the absolute amount of bound ^{3}H-cyclic AMP in the LH samples was more than 5 times that in the control preparations.

CYCLIC AMP DEPENDENT PROTEIN KINASE

Several authors have reported that cyclic AMP binds the regulatory portion of the protein kinase enzyme (7-10) and, in view of this fact, we considered the possibility that this enzyme system might represent part of the macromolecular binding material. Protein kinase assays were carried out on homogen-

128

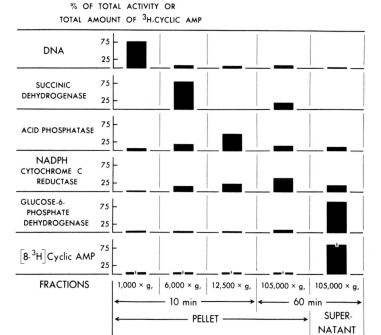

Fig. 2. The subcellular localization of [8-³H] cyclic AMP, DNA and marker enzymes in slices of corpora lutea which were incubated with 10 μg/ml of LH. The height of the bars represent the mean and the heavy line the standard error of the mean. Approximately 1 g of slices of a corpus luteum were incubated in 10 ml of medium as described in the legend to Fig. 1, except that 2.5μCi/ml of [8-³H] adenine was used in the preincubation. The slices were homogenized after the second incubation in 2 ml of 0.25 M sucrose; 0.05 M Tris-HCl, pH 7.4; 0.04 M theophylline; and 0.02 M EDTA at 0⁰. The homogenate was separated into the fractions indicated in the figures. The pellets and the 105,000 x g supernatant were analyzed for DNA, marker enzymes and [8-³H] cyclic AMP.

129

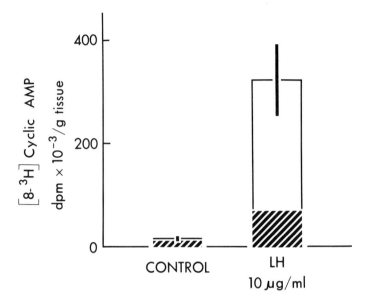

Fig. 3. The measurement of bound [8-^{3}H] cyclic AMP in the cytosol fraction. In 8 experiments, cytosol fractions were prepared from LH (10μg/ml) treated slices. The amount of [8-^{3}H] cyclic AMP bound to a macromolecular component was determined by ultrafiltration using an Amicon PM-30 membrane as described in the text. The height of the bars represent the mean incorporation of [8-^{3}H] adenine into cyclic AMP and the heavy line the standard error of the mean. The shaded portion represents the percent of the [8-^{3}H] cyclic AMP bound to a macromolecule.

ates and the cytosol fractions of corpora lutea, using histone as the substrate, and it was found that this enzyme activity was indeed present in corpora lutea, and that it was largely in the cyclic AMP dependent form. In addition this protein kinase activity was retained by a PM-30 membrane upon ultrafiltration of a cytosol fraction.

The subcellular localization of protein kinase activity was then investigated. Again five subcellular fractions were prepared as before using a homogenization medium consisting of 0.05 M Tris-HCl buffer, pH 7.4, 0.025 M KCl, 0.005 M $MgCl_2$ in 0.25 M sucrose to facilitate the assay of protein kinase.

As with cyclic AMP, the majority of protein kinase activity was present in the cytosol fraction. (Figure 4). The next most active fraction was the 1,000 x g pellet while the other fractions had only minimal activity. The pattern of protein kinase activity supports the possibility that this enzyme might represent some of the macromolecular cyclic AMP binding material.

Cyclic AMP dependent protein kinase was then isolated according to the plan shown in Scheme 1. Further purification of protein kinase was attempted by chromatography of material from peak B of the hydroxylapatite column on a Sephadex G-200 column. The column was eluted with 0.01 M potassium phosphate buffer, pH 7.4, 0.001 M EDTA, and 0.001 M DTT. No purification of protein kinase was achieved on the Sephadex G-200 column, however, and there were large losses of catalytic activity. During the purification procedure different protein substrates were tested to determine if more than one protein kinase enzyme with different substrate specificities could be isolated. Four substrates were used, calf thymus histone (type II-A, Sigma); protamine sulfate (essentially histone free, Sigma); casein; and glycogen synthetase (partially purified from rabbit skeletal muscle).

The highest specific activity of kinase activity was found with protamine as substrate followed by histone and finally casein. There appears to be considerable protein kinase activity with glycogen synthetase as the substrate but a strictly quantitative comparison with other substrates is not yet possible.

The protein kinase activities associated with these different substrates did not seem to purify together as shown by the changes in yield and specific activity during purification as illustrated in Figure 5, A and B. The yield of histone kinase activity increased when the cytosol fraction was prepared and when the ammonium sulfate pellet was made indicating that an inhibitor

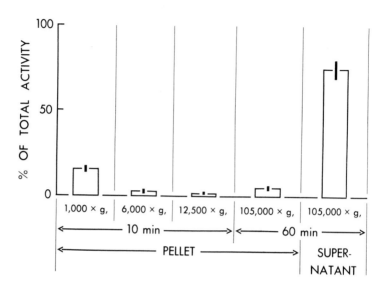

Fig. 4. The subcellular localization of cyclic AMP dependent protein kinase activity in the bovine corpus luteum. The bars represent the percent of the total recovered activity and the heavy vertical lines the standard error of the means. Corpus luteum tissue was homogenized in a 0.25 M sucrose containing 0.005 M Tris-HCl buffer, pH 7.4, 0.025 M KCl, and 0.005 M MgCl$_2$. Subcellular fractions were prepared by differential centrifugation. Protein kinase activity was assayed at pH 6.5 in a reaction mixture containing 10μmoles glycerol-P, 4μmoles NaF, 0.4μmoles theophylline, 0.06μmoles EGTA, 0.04μmoles EDTA, 2.0μmoles magnesium acetate, 0.6mg histone type II-A (Sigma), 10nmoles cyclic AMP, 40nmoles [γ-^{32}P] ATP, and an aliquot of the particular subcellular fraction in a total volume of 0.2ml. The reaction was initiated by the addition of the [γ-^{32}P] ATP. Samples were incubated at 30° for 5 min, after which 50μl aliquots were spotted on circles of Whatman 31 ET filter paper and placed immediately in ice cold 10% TCA to terminate the reaction. The papers were washed successively in cold 10% TCA (30 min), cold 5% TCA (30 min), room temperature 5% TCA (30 min) and 95% ethanol (10 min). The papers were dried and then counted in toluene scintillation fluid to determine protein bound ^{32}P.

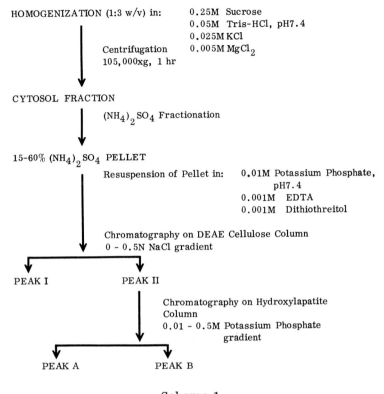

HOMOGENIZATION (1:3 w/v) in: 0.25M Sucrose
                                 0.05M Tris-HCl, pH7.4
                                 0.025M KCl
                                 0.005M MgCl$_2$
            Centrifugation
            105,000xg, 1 hr

CYTOSOL FRACTION

            (NH$_4$)$_2$SO$_4$ Fractionation

15-60% (NH$_4$)$_2$SO$_4$ PELLET
            Resuspension of Pellet in: 0.01M Potassium Phosphate,
                                               pH7.4
                                         0.001M EDTA
                                         0.001M Dithiothreitol

            Chromatography on DEAE Cellulose Column
            0 - 0.5N NaCl gradient

PEAK I   PEAK II

                Chromatography on Hydroxylapatite
                Column
                0.01 - 0.5M Potassium Phosphate
                             gradient

        PEAK A   PEAK B

Scheme 1

Procedure used for preparing cyclic AMP dependent protein kinase from bovine corpus luteum.

might have been separated from the enzyme. The yield of protein kinase activity associated with protamine or casein, on the other hand, fell steadily as the purification was carried out. The final yield of protein kinase was 39% with histone, 10% with protamine, and 4% with casein.

The changes in purification of the enzyme with the different substrates show a similar variability. The greatest purification was achieved with histone as the substrate with a 43 fold increase

133

in specific activity.   The purification of protamine kinase activity
was 12 fold and casein kinase activity only about 4 fold.

One interpretation of these results is that we are not
dealing with one enzyme having a broad substrate specificity
but with multiple protein kinases of differing substrate
specificities. The changes in relative yield and specific
activity might then be due to selective losses of individual
enzymes.

A clear physical separation of the protein kinases associated
with these three substrates has not been achieved using our purification procedure, but a protein kinase activity which acts on
glycogen synthetase does separate from the other protein kinases
during the hydroxylapatite column chromatography step.
(Figure 6).   The upper panel illustrates the elution pattern of

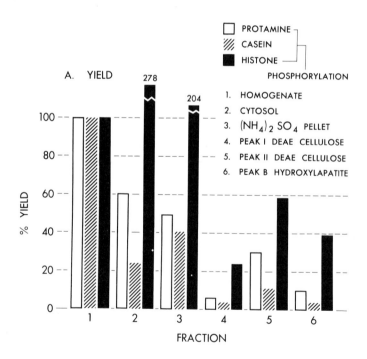

B. PURIFICATION

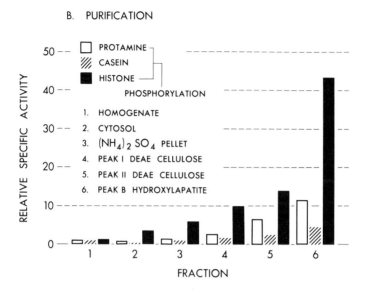

FRACTION

Fig. 5. Yield (A) and purification (B) of cyclic AMP dependent protein kinase activities. Protein kinase was purified according to method in Scheme 1. Fractions from the different stages
of purification were treated for their abilities to phosphorylate
protamine sulfate, casein and histone type II-A. Phosphorylation
of substrates was assayed by the method given in Fig. 4
except that sodium tungstate (0.25%) was used in the TCA washes
when protamine sulfate was the substrate.

A) The yield of protein kinase activities of the homogenate
1. were set equal to 100% and the yields of activities of the other
fractions were then calculated.

B) The three kinase activities of the homogenate were set
equal to 1, and the three relative specific activities of each of the
other fractions were then calculated.

protein and the binding of ^{3}H-cyclic AMP. The middle panel
illustrates the protein kinase activity with protamine and the
lower panel the protein kinase activity associated with glycogen
synthetase.

As seen in the lower panel, there are several peaks of activity with glycogen synthetase. The first peak, which includes the material in fraction 15 to 30, had no activity with protamine, histone, or casein. The elution pattern of protamine kinase shown in the middle panel illustrates the lack of this activity in these early fractions. The elution patterns of histone and casein kinase activities were almost identical to that for protamine. The fractions 30 to 65 have activity with all substrates including glycogen synthetase. Fractions 70 to 80 also had activity with glycogen synthetase with little or no activity with the other substances.

ROLE OF CYCLIC AMP IN STEROIDOGENESIS

At the same time we were purifying the protein kinase, we were attempting to determine if it played any role in the stimulation of steroidogenesis. Fig. 7 illustrates some of the possible sites of cyclic AMP action. First of all it could directly enhance the side-chain cleavage steps involved in the conversion of cholesterol to pregnenolone.(12). Secondly, it might enhance the release of pregnenolone from the mitochondria. This would result in an increase in side-chain cleavage activity since pregnenolone has been reported to be an inhibitor of the cleavage reaction.(13). Thirdly, it might enhance the entrance of the substrate, cholesterol, into the mitochondria, which might be the rate limiting step.(14). Finally, cyclic AMP might stimulate the cholesterol esterase enzyme; this could provide a large pool of free cholesterol. Behrman and Armstrong (15) have reported that the administration of LH to rats in vivo does stimulate this enzyme activity.

We have examined two of these possibilities. The first was to see whether cyclic AMP and/or protein kinase had any direct effect on the cholesterol side-chain cleavage enzyme system. Cyclic AMP and protein kinase were added in various combinations with ATP and magnesium to whole homogenates, mitochondrial preparations, and solubilized preparations of the cleavage system. None of these experiments indicated that cyclic AMP or protein kinase could directly stimulate this reaction.

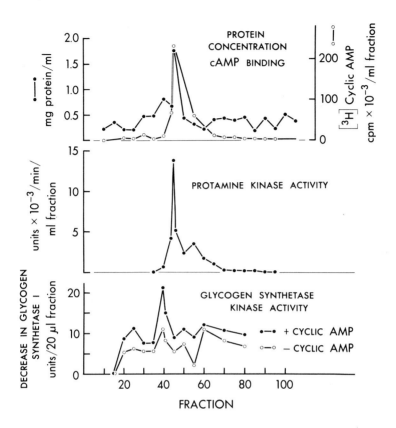

Fig. 6. Hydroxylapatite chromatography of cyclic AMP dependent protein kinase. An aliquot of the peak II DEAE cellulose column eluate was loaded onto a 1.5 x 20 cm hydroxylapatite column and eluted with 0.01 M potassium phosphate, pH 7.4, 0.001 M EDTA and 0.001 M dithiothreitol until fraction 11 when a 0.01 - 0.5 M potassium phosphate gradient was started (total volume of gradient was 625 ml). The eluate was collected in 5ml fractions.

Upper panel: Concentration of protein ($\bullet$ - $\bullet$). ^{3}H - cyclic AMP binding (o - o) to fractions was determined by incubating a 50 μl aliquot of each fraction at pH 4.0 for 1 hour at 0^0 in the presence of 24μmoles sodium acetate, 2μmoles Mg SO$_4$, and 0.5μCi ^{3}H - cyclic AMP (28 Ci/mmole). At the end of the in-

137

cubation, 2ml of 0.02 M potassium phosphate pH 6.0 was added and the mixture filtered through a callulose acetate filter (0.45µ). The filter was washed with an additional 8ml of potassium phosphate, dried and counted in 10ml of Bray's solution.

Middle panel: Total protamine kinase activity was measured as in Fig. 4.

Lower panel: Conversion of synthetase I to glycogen synthetase D. The conversion of glycogen synthetase from its I to D form was measured at 30° and pH7.0 in a reaction mixture of 100µl containing 10mM Tris-HC1, 3mM ATP, 1mM magnesium sulfate, 1x10⁻⁴M cyclic AMP (when added), 50-100µg partially purified glycogen synthetase, and an aliquot of the fraction being tested.

To the appropriate tubes, 200µl of 5mM EDTA and 0.3mM dithiothretol were added before incubation so that initial level of independent and dependent glycogen synthetase activity could be determined. The samples were incubated for 30 min and the reaction terminated by the addition of 200µl of the EDTA dithiothreitol solution. Aliquots (25µl) were removed and assayed for total and independent glycogen synthetase activity according to the method of Thomas et al. (11).

An examination of the cholesterol esterase system was then carried out, since Behrman and others found a stimulation of the enzyme, and the reaction is similar to that of the hormone sensitive lipase.

The first approach was to assess the effect of LH and exogenous cyclic AMP in cholesterol esterase activity in slice incubations. Slices of a corpus luteum were prepared and incubated as usual under control conditions or with LH or cyclic AMP for 1 hour. The slices were then rinsed in isotonic saline and homogenized in phosphate buffer. The homogenate were then assayed for cholesterol esterase activity using cholesterol palmitate-1-¹⁴C as the substrate, according to the procedure of Behrman and Armstrong. (15).

The results of ten incubations are shown in Figure 8. The mean values of the LH treated and cyclic AMP treated tissues

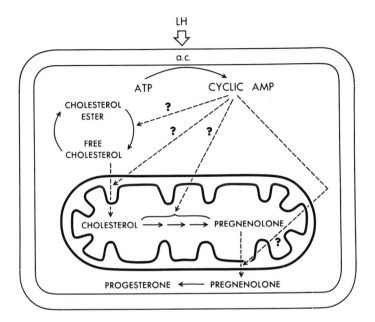

Fig. 7. Possible sites of action of cyclic AMP in mediating the steroidogenic effect of luteinizing hormone in the corpus luteum.

were slightly higher than the controls but the effect was not consistent and the probability that this increase was due to chance alone was one out of ten. These results were therefore uncertain and attempts were then made to demonstrate a direct effect of cyclic AMP and the partially purified protein kinase on relatively crude cholesterol esterase preparations; these included the 105,000 x g supernatant fraction and a 0 - 30% $(NH_4)_2$ SO_4 pellet from the preceeding supernatant. So far, however, neither of these preparations has consistently responded to cyclic AMP or protein kinase.

SUMMARY

We have shown therefore that the endogenous cyclic AMP

produced in the corpus luteum cell is localized predominantly in the cytosol fraction and a significant portion of this is bound to macromolecular material. Some of the macromolecular material appears to have protein kinase activity. On purifying this enzyme activity we have found that there may be multiple protein kinases in the corpus luteum with different specificities rather than one enzyme with multiple specificities. In regard to our goal of understanding the mechanism of action of cyclic AMP on steroidogenesis we have fallen short. So far we have not been able to link protein kinase with any part of steroidogenesis.

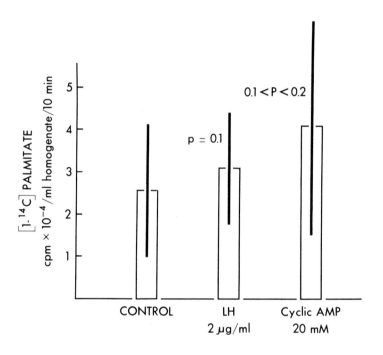

Fig. 8. Cholesterol esterase activity of corpora lutea slices incubated under control conditions or in the presence of luteinizing hormone or cyclic AMP, for details of incubation see reference (1) and for assay see reference (15).

ACKNOWLEDGEMENTS

The authors thank Dr. F. Huijing and Mrs. L. Clerch for providing the rabbit skeletal muscle glycogen synthetase.

REFERENCES

(1)   N. R. Mason, J. M. Marsh and K. Savard, J. Biol. Chem., 237 (1962) 1801.

(2)   J. M. Marsh, R. W. Butcher, K. Savard and E. W. Sutherland, J. Biol. Chem., 241 (1966) 5436.

(3)   J. M. Marsh and K. Savard, Steroids, 8 (1966) 133.

(4)   J. M. Marsh, J. Biol. Chem., 245 (1970) 1596.

(5)   J. F. Kuo and E. C. deRenzo, J. Biol., Chem. 244 (1967) 2252.

(6)   G. Krishna, B. Weiss and B. B. Brodie, J. Pharmacol. Expt. Ther., 163 (1968) 379.

(7)   G. N. Gill and L. D. Garren, Biochem. Biophys. Res. Commun., 39 (1970) 335.

(8)   M. A. Brostrom, E. M. Reimann, D. A. Walsh and E. G. Krebs in: Advances in Enzyme Regulation, Vol. 8, ed. G. Weber (Pergamon Press, New York, 1970) 191.

(9)   M. Tao, M. L. Salas and F. Lipmann, Proc. Nat. Acad. Sci. U.S.A., 67 (1970) 408.

(10) A. Kuman, H. Yamamura and Y. Nishizuka, Biochem. Biophys. Res. Commun., 41 (1970) 1290.

(11) J. A. Thomas, K. K. Schlender and J. Larner, Anal. Biochem., 25 (1968) 486.

(12)   S. Roberts, R. W. McCune, J. E. Creange and P. O.
       Young, Science, 158 (1967) 372.

(13)   S. B. Koritz and P. F. Hall, Biochemistry, 3 (1964) 1298.

(14)   O. Hechter in: Vitamins and Hormones, Vol. 13, ed.
       R. S. Harris, G. F. Marrian and K. V. Thimann (Academ-
       ic Press, New York, 1955) 293.

(15)   H. R. Behrman and D. T. Armstrong, Endocrinology, 85
       (1969) 474.

DISCUSSION

R. SHARMA: Is the rate limiting step in conversion of
cholesterol to pregnenolone the entry of the cholesterol
into the mitochondria?

J.M. MARSH: No, we have no evidence directly for or
against that hypothesis.

R. SHARMA: Do you have any evidence that the rate limiting step perhaps could be the expulsion of pregnenolone from
mitochondria to the cytosol?

J.M. MARSH: No, not from our laboratories. This has been
proposed by others but we do not have evidence for it.

G.N. GILL: I would like to comment on the cholesterol
esterase. I suspect the corpus luteum will turn out to be
very much like the adrenal cortex. Simpson has recently
reported that indeed they could activate the cholesterol
esterase by protein kinase and cyclic AMP in a manner analogous to the hormone sensitive lipase. I think that even
if this event occurs, that this would not explain the rate
limiting step in steroid hormone production, since, if one
blocks steroid production in both of these tissues with
protein synthesis inhibitors, one can still activate the
cholesterol esterase but the free cholesterol just piles
up and cannot go further. It is likely that the rate
limiting step is going to be distal to that.

J.M. MARSH: I do not think that this experiment with protein synthesis inhibitors really indicates what is the rate
limiting step. The cholesterol esterase could be the rate
limiting step before the treatment with protein inhibitors,
but if the inhibitor blocks the steroidogenic pathway,
distal to the esterase the product cholesterol has to pile
up, whether the cholesterol esterase was the slowest step
or not. On the other hand, I do not think there is sufficient evidence yet to pinpoint any of the possible steps
as the rate limiting one.

G.H. DIXON: I was struck by the purification and properties
you observed of your kinase, that it was rather similar to
one that Dr. Bengt Jergil had isolated from trout testis
a couple of years ago and in particular the inactivation
on Sephadex. But the thing I really wanted to ask was that
you observed this rather variable relationship between the
rates of protamine phosphorylation and histone phosphorylation. What salt concentration did you do these experiments at, because this is critical. We found this to be
critical for our trout testis kinase in that in high salt,
you got very rapid phosphorylation of protamine and low
phosphorylation of histone. But with low salt this situation was reversed. When I say high salt I mean about
.3 to .4 molar.

J.M. MARSH: The assays were carried out at low salt concentration.

G.H. DIXON: Have you ever tried it at quite a high salt
concentration?

J.M. MARSH: No, we have not tried that.

E.G. KREBS: I was particularly interested in your ability
to separate a synthetase I kinase activity from the other
protein kinase activity. This is something we have failed
to do in skeletal muscle. Did you use the change in synthetase activity as your assay system?

J.M. MARSH: Yes, that is correct.

E.G. KREBS: Did you do any assays where you utilized protein phosphorylation of the synthetase?

143

J.M. MARSH: No, we did not do that. These assays were done solely on the basis of change from I to D form of the synthetase.

CYCLIC NUCLEOTIDES, PROTEIN PHOSPHORYLATION AND THE

REGULATION OF FUNCTION IN NERVOUS TISSUE

P. GREENGARD
Department of Pharmacology
Yale University School of Medicine

Abstract: It now seems likely that many of the effects of
cyclic AMP are mediated through regulation of the phosphorylation of key cellular proteins by protein kinases.
Nevertheless, in the case of the majority of actions of
cyclic AMP, we still have little or no idea as to the
identity of the substrate protein involved. Our ignor-
ance of the nature of the substrate proteins for cyclic
AMP-dependent protein kinases extends to nervous tissue.
This is hardly surprising, since the role of cyclic AMP
in neural function is only beginning to be understood.
However, recent studies not only indicate a physiological
role for cyclic AMP in neural function, but suggest how
cyclic AMP-dependent protein kinase and its substrate
protein may be involved in mediating this role of the
cyclic nucleotide.
       A variety of data support the following hypothet-
ical role for cyclic AMP in the physiology of synaptic
transmission at certain types of synapses: neurotransmitter, released from presynaptic nerve endings, activates
an adenylate cyclase in the postsynaptic membrane, and,
thereby, causes the accumulation of cyclic AMP in the
postsynaptic neurons; the newly formed cyclic AMP activates a protein kinase leading to the phosphorylation
of a protein constituent of the plasma membrane; this
phosphorylation of the plasma membrane of the postsynaptic neurons causes an alteration in the movement of ions
across the membrane, resulting in a change in membrane
potential of the cells; upon removal of the phosphate

from the membrane protein, by a protein phosphatase present in the synaptic membrane, the membrane potential returns to its initial value. This scheme provides a mechanism by which cyclic AMP, acting as the mediator for the actions of certain types of neurotransmitters, may regulate the membrane potential of certain types of neurons and, thereby, modulate their excitability.

A more limited amount of data suggest the possibility that cyclic GMP mediates muscarinic cholinergic transmission in neural tissue. The effects of cyclic GMP may be mediated through the regulation of cyclic GMP-dependent protein kinase activity.

INTRODUCTION

I would like to review some of the studies from our laboratory concerned with the role and mechanism of action of cyclic AMP in neuronal function. I shall first summarize our work on the role of cyclic AMP in the physiology of the nervous system in order to provide the appropriate background for presenting our studies concerning the mechanism of action of cyclic AMP in neural tissue.

At the time we began our studies on the role of cyclic AMP in neural function, there was a variety of indirect evidence suggesting a possible involvement of cyclic AMP in the functioning of the nervous system, but no indication as to what this role might be. For instance, it had been found that adenylate cyclase (1) and phosphodiesterase (2) were present in higher concentrations in nervous tissue than in any other tissue which had been examined. Moreover, the central nervous system stimulants, theophylline and caffeine, had been shown to be phosphodiesterase inhibitors (2). In addition, a variety of chemical and physical changes had been shown to lead to altered levels of cyclic AMP in brain tissue (3-8).

PHYSIOLOGICAL STUDIES

In our initial studies on the role of cyclic AMP in neuronal function, we utilized the superior cervical ganglion of the mammalian peripheral autonomic nervous system. Until recently, this ganglion was thought to be composed of a simple monosynaptic cholinergic pathway. However, studies in a

variety of laboratories over the past ten years have shown
that there are interneurons in this ganglion which are inner-
vated by presynaptic cholinergic fibers and which, in turn,
innervate the postganglionic neurons. These interneurons
contain dopamine. When one stimulates the preganglionic
nerve fibers to the superior cervical ganglion and records
from the postganglionic neurons, one observes characteristic
electrical signs of nervous activity, including a compound
action potential, followed by a hyperpolarization, the latter frequently being referred to as a slow inhibitory postsynaptic potential (S-IPSP). The compound action potential
results from activity in the direct monosynaptic (nicotinic)
cholinergic excitatory pathway, and it is thought that the
S-IPSP is the result of dopamine released from the interneurons acting on the postganglionic neurons. Part of the evidence in support of this scheme is the observation of Libet
(9,10), confirmed in our laboratory (11,12), that exogenous-
ly applied dopamine causes a hyperpolarization of the ganglion cells.

The superior cervical sympathetic ganglion seemed to us
an interesting preparation with which to study the possible
role of cyclic AMP in the nervous system, because it is complex enough to have certain integrative properties, yet simple enough, hopefully, to permit analysis. Various aspects
of the studies I am going to describe on the role of cyclic
AMP in the nervous system were carried out in our laboratory
by Drs. D. A. McAfee, M. Schorderet, and J. W. Kebabian.
Among the initial experiments which we carried out, we examined the effect of physiological activity, induced by stimulation of the preganglionic fibers, on the level of cyclic
AMP in the superior cervical ganglion of the rabbit. The
results of stimulating the ganglionic fibers of this preparation, at a physiological frequency, on the level of cyclic
AMP in the ganglion are shown in Figure 1. It can be seen
that the level of cyclic AMP increased by 4 to 5 fold within
one minute of stimulation, and that this elevated level was
maintained for as long as the stimulation was continued.

The results shown in Figure 1 demonstrated that physiological activity in neuronal tissue can result in an increase
in the level of cyclic AMP. The results raised the question
as to whether this increase in cyclic AMP occured in the
axons in association with impulse conduction, or in some

147

ganglionic structure in association with synaptic transmission. The results of one type of experiment designed to answer this question are shown in Figure 2, in which the effect of activity on the level of cyclic AMP is presented for three different types of peripheral nervous tissue, the superior

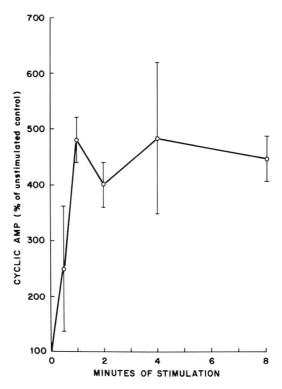

Fig. 1. Increase in cyclic AMP level associated with synaptic activity in rabbit superior cervical ganglia in vitro. Ganglia were stimulated at a frequency of 10 per sec for the indicated periods of time and, immediately following stimulation, were extracted for assay of cyclic AMP. Cyclic AMP levels are expressed as percentage (mean ± standard error) of the values found in non-stimulated contralateral control ganglia. Temperature, 33°C. Four to 11 rabbits for each point. Figures 1-3 are based on data taken from McAfee et al. (13).

cervical ganglion, which contains nerve cell bodies as well
as synapses, the vagus nerve, which contains neither nerve
cell bodies nor synapses, and the nodose ganglion, which
contains nerve cell bodies but not synapses. It can be seen
that only in the superior cervical ganglion was activity
associated with an increase in cyclic AMP level. These results suggested that the increase in cyclic AMP was associated with some process of synaptic transmission. This conclusion was supported by comparing the effect on cyclic AMP
levels of stimulating the superior cervical ganglion through
its presynaptic and through its postsynaptic fibers. In
contrast to the effectiveness of preganglionic stimulation
(which results in synaptic transmission) in increasing the
cyclic AMP level, stimulation of the ganglion through its

EFFECT OF STIMULATION ON CYCLIC AMP
CONTENT OF VARIOUS RABBIT PERIPHERAL
NERVOUS TISSUES

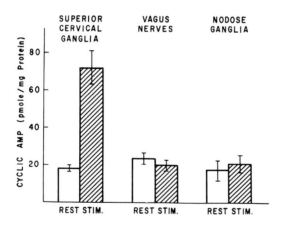

Fig. 2. Effect of stimulation on cyclic AMP
content of isolated rabbit peripheral nervous
tissues. Unshaded bars: cyclic AMP content in
unstimulated control tissue. Shaded bars: cyclic
AMP content in tissue immediately following supra-
maximal stimuli at 10 per sec for 2 min. Temperature, 33°C. Values represent mean ± standard
error for 5-11 rabbits.

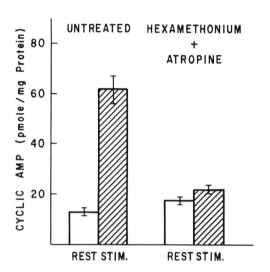

EFFECT OF CHOLINERGIC BLOCKADE
ON CYCLIC AMP CONTENT OF
STIMULATED RABBIT SUPERIOR
CERVICAL GANGLIA

Fig. 3. Effect of cholinergic blockade on cyclic AMP content of stimulated rabbit superior cervical ganglia. Conditions are as in Fig. 2 except that stimuli were applied for only 1 min. The treated tissue was preincubated for 30 min in 1 mM hexamethonium chloride and 0.1 mM atropine sulfate. Values represent mean ± standard error for four to five rabbits.

postganglionic fibers (which does not result in synaptic transmission) did not result in an increased level of cyclic AMP. Thus, these results also indicated that synaptic transmission was required for the increase in cyclic AMP to be manifested.

We next turned to the question as to whether this increase

in cyclic AMP occurred in the presynaptic nerve terminals in association with the process of release of acetylcholine, or in some postsynaptic structure in response to the released acetylcholine. To answer this question, we studied the effect of the cholinergic blocking agents, hexamethonium and atropine, on the increase in cyclic AMP observed in response to preganglionic stimulation. It is generally a- greed that these blocking drugs do not interfere with the release of acetylcholine from the presynaptic terminals, but rather that they inhibit the activation by acetylcholine of postsynaptic acetylcholine receptors. As can be seen in Figure 3, a combination of these two drugs prevented the usual increase in cyclic AMP, indicating that this increase occurred in some postsynaptic structure in response to acetyl- choline released from the presynaptic terminals. There were still three distinct mechanisms by which the observed increase in cyclic AMP might have occurred. The increase in cyclic AMP might have occurred (a) in the postganglionic neurons in response to acetylcholine released via the direct excitatory pathway, (b) in the interneurons in response to the presynaptic release of acetylcholine, or (c) in the post- ganglionic neurons in response to dopamine released from the interneurons. A variety of experimental evidence indicated that the latter possibility was most probably the correct one. One such piece of evidence was the observation that low concentrations of dopamine caused a large increase in the level of cyclic AMP in slices of mammalian superior cervical ganglion (Figure 4). Norepinephrine also caused an increase in the cyclic AMP content of the ganglion, but the concentration of norepinephrine required to produce this effect was much greater than that of dopamine. In order to determine whether the increase in cyclic AMP in response to dopamine occurred through stimulation of a dopamine-sensitive adenylate cyclase or through inhibition of a phosphodiester- ase, experiments were carried out with homogenates of the ganglion (14). In such experiments, it was shown that the increase in cyclic AMP was due to stimulation of a dopamine- sensitive adenylate cyclase; dopamine had no apparent effect on the phosphodiesterase activity of the ganglion homogenate.

The fact that stimulation of the preganglionic nerve fibers or the application of exogenous dopamine each caused an increase in the level of cyclic AMP in the ganglion and also caused a hyperpolarization of the postganglionic neur-

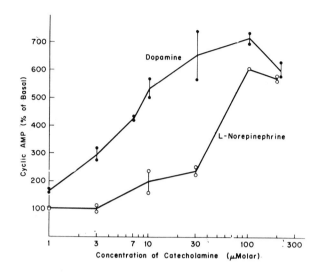

Fig. 4. Stimulation by dopamine and by
l-norepinephrine of cyclic AMP accumulation in
blocks of bovine superior cervical ganglion.
Tissue was incubated for 5 min at 37°C in oxygen-
ated Krebs-Ringer bicarbonate buffer, pH 7.4,
containing 10 mM theophylline and the indicated
amount of catecholamine. Cyclic AMP accumulation
during incubation in the presence of catecholamine
and theophylline is expressed as the percentage
of that observed during incubation in the presence
of theophylline alone. The curves are drawn
through the average of two data points; each
data point represents the mean of duplicate deter-
minations on an individual sample. Figure taken
from Kebabian and Greengard (14).

ons made it important to examine the possibility of a causal
relationship between the increase in cyclic AMP and the
hyperpolarization. It seemed possible that cyclic AMP might
be responsible for mediating the hyperpolarizing action of
dopamine. One experiment indicating that this may be the
case is illustrated in Figure 5, where it can be seen that

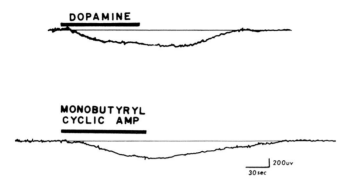

Fig. 5. Effect of dopamine and of monobutyryl cyclic AMP on resting membrane potentials recorded from postganglionic neurons of superior cervical sympathetic ganglion of the rabbit. The duration of superfusion with Locke solution containing dopamine or monobutyryl cyclic AMP is indicated by the solid bars. All records are d-c recording, hyperpolarization downward. Modified from McAfee and Greengard (11).

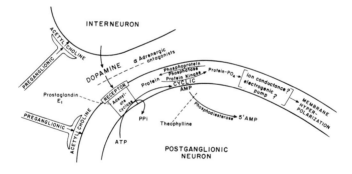

Fig. 6. Model of the proposed mechanism by which cyclic AMP acts as the mediator of dopaminergic transmission and modulator of cholinergic transmission in ganglia of the sympathetic nervous system. Figure taken from Greengard et al. (15).

153

the hyperpolarizing action of dopamine on the postganglionic
nerve cells could be mimicked by the application of monobutyryl cyclic AMP.

A model of our working hypothesis concerning the role
and mechanism of action of cyclic AMP in the functioning of
the superior cervical ganglion is shown in Figure 6. Activity in the preganglionic nerve endings, through a direct excitatory nicotinic cholinergic pathway, causes the depolarization of the postganglionic neurons, leading to the propa-
gation of impulses down the axons of these postganglionic
neurons. At the same time, activity in other preganglionic
nerve endings causes the excitation of the interneurons,
leading to the release of dopamine. The released dopamine
then activates a dopamine-sensitive adenylate cyclase in
the postganglionic neurons, leading to the production of
cyclic AMP in the postganglionic neurons. This newly formed
cyclic AMP causes the activation of a cyclic AMP-dependent
protein kinase present in the postsynaptic portion of the
synaptic membrane. The activated protein kinase causes the
phosphorylation of an endogenous membrane protein, leading
to an alteration of ion transport across the membrane with
a resultant hyperpolarization of the postganglionic plasma
membrane. As a result of the hyperpolarization of the
postganglionic neuron, it becomes less responsive to subsequent impulses in the direct excitatory pathway. Since the
S-IPSP is a transient phenomenon, a mechanism must be available for terminating the hyperpolarization. This is achieved by a protein phosphatase, which converts the phosphorylated form of the membrane protein back to its non-phosphorylated form. According to this model, cyclic AMP mediates
dopaminergic transmission in the ganglion and thereby modu-
lates nicotinic cholinergic transmission.

A variety of experimental data, in addition to that which
I have already summarized, supports this working model for
the role and mechanism of action of cyclic AMP in the functioning of the ganglion. However, in view of the theme of
this symposium, I shall not review any more of our physiological data concerned with this question, but shall turn
to a summary of our studies of protein phosphorylation and
its possible involvement in the functioning of the nervous
system.

PROTEIN PHOSPHORYLATION STUDIES

Following the discovery by Walsh, Perkins and Krebs (16) of a cyclic AMP-dependent protein kinase in skeletal muscle, Miyamoto, Kuo and I looked for and found a similar enzyme in mammalian nervous tissue (17). We purified the enzyme extensively and studied many of its properties. However, rather than discussing that enzyme, which was a soluble protein kinase, at this point, I am going to describe recent studies in our laboratory concerning the protein kinase system in synaptic membrane fractions of neural tissue. These studies were carried out by Drs. Hiroo Maeno and Edward Johnson, and more recently by Dr. Tetsufumi Ueda. Using protamine and histone as substrates, the subcellular distribution of cyclic AMP-dependent protein kinase in nervous tissue was examined. It was found (18) that those subcellular fractions rich in synaptic membranes were enriched in cyclic AMP-dependent protein kinase. We next studied the subcellular distribution of proteins able to act as substrates for cyclic AMP-dependent protein kinase. For that purpose, we used a partially purified brain protein kinase prepared by the procedure of Miyamoto et al. (17). Using this partially purified protein kinase as the enzyme, and studying the ability of various subcellular fractions of brain to serve as substrate, it was found (19) that synaptic membranes were particularly effective as substrates for this enzyme. When the data were expressed as μmoles of phosphate incorporated per mg of membrane protein, the synaptic membranes were found to be approximately as effective as highly purified histones, the most effective substrates which had yet been found, as substrate for the protein kinase. By using histone phosphate and protamine phosphate as substrates, it was also found (20) that synaptic membrane fractions contained high levels of a protein phosphatase, capable of removing phosphate from protein which had been phosphorylated by cyclic AMP-dependent protein kinase. Thus, the synaptic membrane fractions contained all three protein components of the protein kinase system, namely a protein kinase, a substrate protein, and a protein phosphatase. The ability of these three endogenous protein components of the synaptic membrane to interact with one another is illustrated in the experiment of Figure 7. When synaptic membranes were incubated in the presence of $[\gamma-^{32}P]ATP$ and cyclic AMP, an endogenously catalyzed phosphorylation of

155

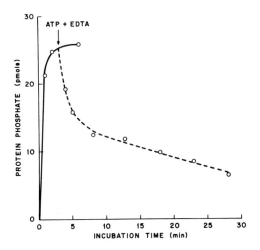

Fig. 7. Incorporation of phosphate into, and
release of phosphate from, synaptic membrane protein. The complete reaction mixture contained
5 μmoles of sodium acetate buffer, pH 6.4, 30 mμmoles
of ethylene glycol bis (β-aminoethyl ether)-N,N'-
tetraacetic acid, 1 μmole of magnesium acetate,
0.5 mμmoles of cyclic AMP, 0.63 mμmoles of [γ-^{32}P]ATP
and 12.5 μg of synaptic membrane protein in a total
volume of 0.1 ml. The reaction was initiated by
the addition [γ-^{32}P]ATP and carried out at 30°.
At the point indicated by the arrow, 0.2 ml of a
solution containing 0.78 mM cold ATP and 7.4 mM
EDTA was added. Figure taken from Maeno and Greengard (20).

synaptic membrane protein was observed, demonstrating the
ability of endogenous protein kinase to phosphorylate an
endogenous substrate protein. After five minutes of incubation, a large amount of cold ATP, together with EDTA, was
added to the reaction mixture to terminate the incorporation
of radioactive phosphate into the membrane protein. The
cold ATP greatly lowered the specific activity of the radio-

active ATP, and thereby decreased ^{32}P incorporation; the EDTA, as shown in separate experiments, strongly inhibited the protein kinase reaction. Immediately upon the addition of the cold ATP and EDTA, there ensued a biphasic disappearance of ^{32}P from the membrane protein, indicating an interaction between the phosphorylated membrane protein and an endogenous protein phosphatase.

In view of the fact that the synaptic membrane is composed of many different types of proteins, it seemed of importance to determine whether the effectiveness of these synaptic membranes as substrates for cyclic AMP-dependent protein kinase might be attributable to incorporation of the phosphate into only one or just a few of the membrane proteins. For this reason, we incubated synaptic membranes with [γ-^{32}P]ATP in the absence and presence of cyclic AMP and, after two minutes of incubation, terminated the reaction by the addition of sodium dodecyl sulphate. The sodium dodecyl sulphate, in addition to terminating the reaction, also caused the solubilization of the membrane protein. We then subjected the solubilized membrane protein to polyacrylamide gel electrophoresis and examined the pattern of ^{32}P incorporation into the various membrane proteins by autoradiography (21). Under the conditions used, several proteins became phosphorylated, but the phosphorylation of only one of these proteins was dramatically affected by cyclic AMP. This particular protein represented a minor component of the synaptic membrane, as judged by the protein staining pattern of the gels. We are currently engaged in efforts to further characterize the cyclic AMP-dependent protein kinase system of synaptic membranes, and particularly to compare the physico-chemical properties of the substrate protein in the phosphorylated and non-phosphorylated state, in order to determine whether the phosphorylation of this membrane protein is associated with a configurational change. The results of such studies could be quite interesting, since we believe this protein may be involved in some of the ion permeability changes associated with the physiology of synaptic membranes. We also intend to compare the properties of the protein kinase system of synaptic membranes with the properties of the soluble cyclic AMP-dependent protein kinase system from brain.

I would like to turn now to certain properties of the

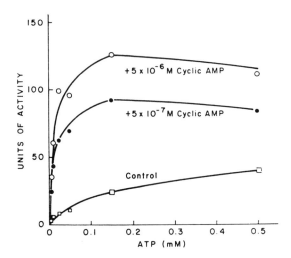

Fig. 8. Effect of ATP concentration on the activity of a partially purified cyclic AMP-dependent protein kinase from bovine brain, in the absence and presence of cyclic AMP. Activities have been corrected for zero time values determined at each ATP concentration. Figure taken from Miyamoto et al. (17).

soluble protein kinase system from mammalian brain. In general, the properties of the protein kinases obtained from various tissues have proven to be similar (22). However, in at least one respect, the soluble brain protein kinase which we have studied (17) is different from those of most other tissues. This difference concerns the effect of cyclic AMP on the apparent affinity of the enzyme for ATP. With the protein kinases derived from most tissues, activation of the enzyme by cyclic AMP is associated primarily with an increase in the V_{max}, with a relatively slight effect on the apparent affinity of the enzyme for ATP (22). In the case of the partially purified protein kinase from bovine brain (17,22), cyclic AMP had little effect on the V_{max}, but caused a dramatic change in the affinity of the enzyme for ATP (Figure 8).

Studies in several laboratories (23-28) have indicated that cyclic AMP-dependent protein kinases in a number of tissues are composed of regulatory and catalytic subunits. We have confirmed that a similar situation occurs in the case of protein kinase from nervous tissue. Thus, as with the enzyme from other tissues, cyclic AMP-dependent protein kinase from bovine brain is composed of regulatory and catalytic subunits, and cyclic AMP appears to act by removing the regulatory subunit from the holoenzyme, leaving the catalytic subunit free to catalyze the protein kinase reaction (29,30). In addition, it has been found (29,30) that various model substrate proteins are also capable of bringing about the dissociation of the brain cyclic AMP-dependent protein kinase into regulatory and catalytic subunits, thereby converting the protein kinase into a cyclic AMP-independent form (Figure 9). This effect has recently been confirmed by Tao (31) using a cyclic AMP-dependent protein kinase from rabbit erythrocytes. It has been speculated (29,30) that the dissociation and activation of protein kinase holoenzymes by substrate proteins may constitute an important physiological control mechanism. Thus, it seems possible that regulation of the level of protein kinase activity by substrate proteins may constitute an intracellular regulatory mechanism, analogous to the regulation of these enzymes by cyclic AMP in response to extracellular signals.

CYCLIC GMP STUDIES

Recently, we have obtained evidence suggesting a role and mechanism of action of cyclic GMP in neural tissue. Studies in our laboratory with mammalian brain, heart muscle and intestinal smooth muscle, had suggested the generalization that interaction of acetylcholine with muscarinic receptors, but not with nicotinic receptors, causes an increase in the level of cyclic GMP (32). Subsequently, we obtained evidence indicating that cyclic GMP may mediate synaptic transmission at muscarinic cholinergic synapses of the mammalian superior cervical ganglion. Thus, activation of muscarinic cholinergic receptors in the ganglion, which causes a depolarization of the postsynaptic membrane (33), also causes an increase in the level of cyclic GMP in the ganglion (34). In addition, low concentrations of dibutyryl cyclic GMP mimic the acetylcholine-induced depolarization of the postganglionic neurons of the sympathetic ganglion (11).

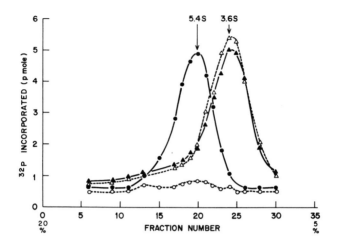

Fig. 9. Dissociation of bovine brain protein
kinase IB by histone. The enzyme (20 μg) was preincubated at 30° for 5 min in a volume of 0.22 ml
containing 50 mM acetate buffer (pH 6.0), 0.3 mM
EGTA, and 2.5 mM 2-mercaptoethanol, in the absence
(o,●) or presence (Δ,▲) of histone (1 mg per ml).
At the end of this preincubation, the entire volume of the solution was layered onto 4.8 ml of a
5 to 20% sucrose density gradient containing the
same concentrations of acetate buffer, EGTA, and
2-mercaptoethanol, with or without histone, as were
present in the preincubation solution. After centrifugation, the fractions obtained from each tube
were assayed for protein kinase activity in the
absence (····) or presence (——) of 5 μM cyclic
AMP. Figure taken from Miyamoto et al. (30).

Thus, it seems possible that, in the ganglion, the hyperpolarizing action of dopamine is mediated through increased
levels of cyclic AMP, and the depolarizing action of acetylcholine at muscarinic receptors is mediated through increased levels of cyclic GMP.

Concerning the mechanism by which cyclic GMP might mediate the neurotransmitter action of acetylcholine at muscarinic synapses, we have found a family of cyclic GMP-dependent protein kinases in nature (35-37). These cyclic GMP-dependent protein kinases appear to have a fairly widespread distribution. They have been found in mammalian brain, uterus, and bladder (36), as well as in a variety of invertebrate tissues (35-37). It has been difficult to work with the mammalian cyclic GMP-dependent protein kinases because of the relatively low activity of these enzymes compared with that of the cyclic AMP-dependent protein kinases present in the same tissues. [The recent demonstration (38) that phosphate buffer enhances the activity of cyclic GMP-dependent protein kinase and inhibits the activity of cyclic AMP-dependent protein kinase in homogenates of mammalian cerebellum may facilitate the study of the mammalian cyclic GMP-dependent protein kinases.] For this reason, most of our studies of cyclic GMP-dependent protein kinases have been carried out using enzymes from invertebrate sources. One good source of cyclic GMP-dependent protein kinase is lobster tail muscle, extracts of which contain significant amounts of both cyclic AMP-dependent and cyclic GMP-dependent protein kinase activity. These two types of enzyme activity, both present in a partially purified preparation from lobster tail muscle, can be readily separated from each other by chromatography on DEAE cellulose. The cyclic nucleotide dependence of the cyclic GMP-dependent protein kinase is shown in Figure 10, where it can be seen that cyclic GMP was considerably more effective than cyclic AMP in activating this enzyme.

We have studied the mechanism by which cyclic GMP brings about the activation of cyclic GMP-dependent protein kinases. The mechanism appears to be similar to that by which cyclic AMP brings about the activation of cyclic AMP-dependent protein kinases. Thus, it would seem that the cyclic GMP-dependent protein kinases are also composed of regulatory and catalytic subunits, with the regulatory subunits maintaining the enzyme in an inhibited state; cyclic GMP, by removing the regulatory subunit, would appear to unmask the activity of the catalytic subunit. Moreover, as in the case of the cyclic AMP-dependent protein kinases, substrate proteins are able to cause the partial dissociation and activation of cyclic GMP-dependent protein kinases.

161

The effect of histone and of cyclic GMP on the sedimentation rate and on the activity of the cyclic GMP-dependent protein kinase from lobster tail muscle is shown in Figure 11. Preincubation with histone, followed by sucrose density gradient centrifugation, caused the partial dissociation of the holoenzyme, which had a sedimentation coefficient of 7.7 S, into a cyclic GMP-independent catalytic subunit with a sedimentation coefficient of 3.6 S. Similarly, preincubation in the presence of cyclic GMP caused the partial dissociation of the 7.7 S holoenzyme with the appearance of a 3.6 S catalytic subunit. When both histone and cyclic GMP were present in the preincubation medium, there was total dissociation of the holoenzyme into the 3.6 S catalytic subunit. We have not yet found conditions under which preincubation with histone alone or with cyclic GMP alone was able to cause the total dissociation of the holoenzyme and it is

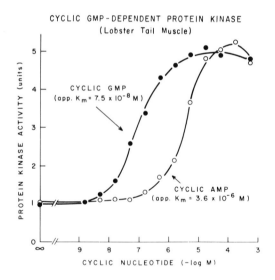

Fig. 10. Effect of varying concentrations of cyclic GMP and cyclic AMP on the activity of a partially purified cyclic GMP-dependent protein kinase from lobster tail muscle. Figure modified from Kuo and Greengard (35).

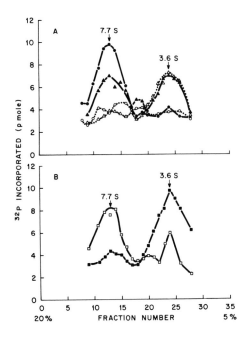

Fig. 11. Dissociation of lobster muscle cyclic
GMP-dependent protein kinase by histone and cyclic
GMP. The enzyme (2.0 mg) was preincubated at 30°
for 5 min in a volume of 0.22 ml containing 50 mM
sodium acetate buffer (pH 6.0), 0.3 mM EGTA, 2.5
mM 2-mercaptoethanol, and the following additions:
(A) none (o,•), 1 mg of histone per ml (Δ,▲); (B)
50 μM cyclic GMP (□), or 1 mg of histone per ml
plus 50 μM cyclic GMP (■). At the end of the preincubation, the entire volume of the solution was
layered onto 4.8 ml of a 5 to 20% sucrose density
gradient containing the same concentrations of
acetate buffer, EGTA, and 2-mercaptoethanol, and
the same additions as were present in the preincubation solution. After centrifugation, the fractions obtained from each tube were assayed for
protein kinase activity in the absence (····)
or presence (——) of 5 μM cyclic GMP. Figure
taken from Miyamoto et al. (30).

163

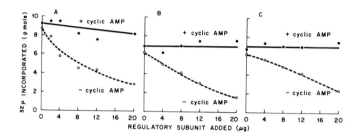

Fig. 12. Inhibition of the activity of isolated catalytic subunits from cyclic AMP-dependent and cyclic GMP-dependent protein kinases, by the addition of regulatory subunit from a cyclic AMP-dependent protein kinase, and restoration of cyclic AMP dependence in the reconstituted holoenzyme. Various amounts of isolated regulatory subunit prepared from bovine brain cyclic AMP-dependent protein kinase were added to (A) catalytic subunit (1.6 μg) from bovine brain cyclic AMP-dependent protein kinase II, (B) catalytic subunit (2.4 μg) from bovine brain cyclic AMP-dependent protein kinase IC, or (C) catalytic subunit (1.6 μg) from lobster muscle cyclic GMP-dependent protein kinase. Enzyme activity was assayed in the absence (o----o) and presence (●——●) of 5 μM cyclic AMP. Figure taken from Miyamoto et al. (30).

conceivable that in vivo both the substrate protein and the cyclic nucleotide may be necessary to achieve the total dissociation of the cyclic GMP-dependent protein kinase.

It seemed of interest to study the possible interaction of subunits obtained from the cyclic AMP-dependent protein

kinase of mammalian brain with the subunits obtained from
the cyclic GMP-dependent protein kinase of lobster tail
muscle. The results of such an experiment are shown in
Figure 12. It was found that the regulatory subunit prepared from bovine brain cyclic AMP-dependent protein kinase
was able to inhibit the activity not only of catalytic subunits prepared from bovine brain cyclic AMP-dependent protein
kinase, but also the activity of catalytic subunits isolated
from the cyclic GMP-dependent protein kinase of lobster tail
muscle. The inhibitory action of the regulatory subunit of
the bovine brain enzyme could be prevented by the addition
of cyclic AMP to the incubation medium. The fact that the
regulatory subunit of bovine brain cyclic AMP-dependent
protein kinase was able to inhibit the catalytic activity
of lobster muscle cyclic GMP-dependent protein kinase suggests that the interactions between regulatory and catalytic subunits have a relatively low degree of specificity,
since these subunits were prepared from different classes
(i.e., cyclic AMP-dependent and cyclic GMP-dependent) of
protein kinase, from different tissues (brain and muscle),
and from different phyla (vertebrates and arthropods).
The low specificity of the interaction between regulatory
and catalytic subunits indicated by these results suggests
that, within those individual cells which contain multiple
protein kinases, there may exist special mechanisms to in-
sure that regulatory and catalytic subunits become associated with their appropriate partners.

One interesting question raised by the experiment of
Figure 12 concerns the cyclic nucleotide specificity of the
hybrid enzyme composed of a regulatory subunit prepared from
a cyclic AMP-dependent protein kinase and a catalytic subunit prepared from a cyclic GMP-dependent protein kinase.
As shown in Figure 13, cyclic AMP was more effective than
cyclic GMP in bringing about the activation of this hybrid
enzyme. Thus, as might have been predicted, the cyclic
nucleotide specificity of the hybrid enzyme is determined
by the regulatory subunit, rather than by the catalytic
subunit.

This presentation has been confined to a review of studies from our own laboratory on the role of cyclic AMP in
neuronal function. However, before concluding, I would
like at least to refer to the interesting studies of Bloom,

165

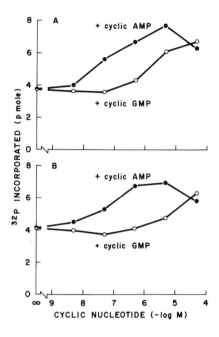

Fig. 13. Cyclic nucleotide dependence of reconstituted protein kinases. Regulatory subunit (20 μg), prepared from bovine brain cyclic AMP-dependent protein kinase, was added to (A) catalytic subunit (2.4 μg) from bovine brain cyclic AMP-dependent protein kinase II, and (B) catalytic subunit (2.0 μg) from lobster muscle cyclic GMP-dependent protein kinase. The enzyme was assayed in the presence of various concentrations of cyclic AMP (●————●) or cyclic GMP (o————o), as indicated. The incorporation of ^{32}P, using the isolated catalytic subunits from the cyclic AMP-dependent and the cyclic GMP-dependent enzymes, was 7.7 and 7.9 pmoles, respectively. Figure taken from Miyamoto et al. (30).

166

Hoffer and Siggins on the Purkinje cells of the mammalian cerebellum. Those investigators have obtained evidence indicating that noradrenergic fibers, originating in the locus coeruleus and terminating in the cerebellum, exert an inhibitory influence on the spontaneous firing of the cerebellar Purkinje cells and that this inhibitory effect of norepinephrine is mediated through cyclic AMP (39,40).

CONCLUSION

In summary, the following comments may be made concerning the role and mechanism of action of cyclic nucleotides in neural tissue. The available evidence suggests that cyclic AMP mediates the action of dopamine at dopaminergic synapses within the superior cervical ganglion, thereby modulating nicotinic cholinergic transmission through the ganglion. In addition cyclic AMP appears to mediate noradrenergic transmission at cerebellar Purkinje cells. We believe that cyclic AMP may achieve these effects through regulating the activity of a protein kinase, the substrate for which may be a specific protein present in the postsynaptic membrane which, by controlling ion transport across the membrane, regulates the potential across this membrane. The available evidence also suggests that cyclic GMP may mediate synaptic transmission at muscarinic cholinergic synapses, and that it may achieve this effect through regulating the activity of a cyclic GMP-dependent protein kinase. We do not yet have any information about the substrate for this cyclic GMP-dependent protein kinase; we are currently attempting to demonstrate the existence of endogenous substrate proteins for cyclic GMP-dependent protein kinases in neural tissue.

A few years ago, we proposed (41,42,35) a general mechanism for the action of cyclic AMP and cyclic GMP in biological tissues. According to that proposal, the diverse effects of cyclic AMP and cyclic GMP are mediated through regulation of the activity of cyclic AMP-dependent and cyclic GMP-dependent protein kinases. The mediation of transmission at certain types of neuronal synapses, through regulation of the levels of cyclic AMP and cyclic GMP, and the consequent alteration of the activity of their respective protein kinases, would represent specific examples of this postulated general mechanism.

REFERENCES

(1) E.W. Sutherland, T.W. Rall and T. Menon, J. Biol. Chem., 237 (1962) 1200.

(2) R.W. Butcher and E.W. Sutherland, J. Biol. Chem., 237 (1962) 1244.

(3) F. Murad, Y.-M. Chi, T.W. Rall and E.W. Sutherland, J. Biol. Chem., 237 (1962) 1233.

(4) L.M. Klainer, Y.-M. Chi, S.L. Freidberg, T.W. Rall and E.W. Sutherland, J. Biol. Chem., 237 (1962) 1239.

(5) B.McL. Breckenridge, Proc. Nat. Acad. Sci. U.S.A., 52 (1964) 1580.

(6) S. Kakiuchi and T.W. Rall, Mol. Pharmacol., 4 (1968) 367.

(7) S. Kakiuchi and T.W. Rall, Mol. Pharmacol., 4 (1968) 379.

(8) S. Kakiuchi, T.W. Rall and H. McIlwain, J. Neurochem., 16 (1969) 485.

(9) B. Libet and T. Tosaka, Proc. Nat. Acad. Sci. U.S.A., 67 (1970) 667.

(10) B. Libet, Fed. Proc., 29 (1970) 1945.

(11) D.A. McAfee and P. Greengard, Science, 178 (1972) 310.

(12) P. Greengard and D.A. McAfee, in: British Biochemical Society Symposia, Vol. 36 (The Biochemical Society, London, 1972) p. 87.

(13) D.A. McAfee, M. Schorderet and P. Greengard, Science, 171 (1971) 1156.

(14) J.W. Kebabian and P. Greengard, Science, 174 (1971) 1346.

(15) P. Greengard, D.A. McAfee and J.W. Kebabian, in:

Advances in Cyclic Nucleotide Research, Vol. 1, eds.
P. Greengard, R. Paoletti and G.A. Robison (Raven
Press, New York, 1972) p. 337.

(16) D.A. Walsh, J.P. Perkins and E.G. Krebs, J. Biol. Chem.,
243 (1968) 3763.

(17) E. Miyamoto, J.F. Kuo and P. Greengard, J. Biol. Chem.,
244 (1969) 6395.

(18) H. Maeno, E.M. Johnson and P. Greengard, J. Biol. Chem.,
246 (1971) 134.

(19) E.M. Johnson, H. Maeno and P. Greengard, J. Biol. Chem.,
246 (1971) 7731.

(20) H. Maeno and P. Greengard, J. Biol. Chem., 247 (1972)
3269.

(21) E.M. Johnson, T. Ueda, H. Maeno and P. Greengard, J.
Biol. Chem., 257 (1972) 5650.

(22) J.F. Kuo, B.K. Krueger, J.R. Sanes and P. Greengard,
Biochim. Biophys. Acta, 212 (1970) 79.

(23) M.A. Brostrom, E.M. Reimann, D.A. Walsh and E.G. Krebs,
in: Advances in Enzyme Regulation, Vol. 8, ed. G. Weber
(Pergamon Press, New York, 1970) p. 191.

(24) G.N. Gill and L.D. Garren, Biochem. Biophys. Res. Comm.,
39 (1970) 335.

(25) M. Tao, M.L. Salas and F. Lipmann, Proc. Nat. Acad.
Sci. U.S.A., 67 (1970) 408.

(26) A. Kumon, H. Yamamura and Y. Nishizuka, Biochem. Biophys. Res. Comm., 41 (1970) 1290.

(27) E.M. Reimann, C.O. Brostrom, J.D. Corbin, C.A. King and
E.G. Krebs, Biochem. Biophys. Res. Comm., 42 (1971) 187.

(28) J. Erlichman, A.H. Hirsch and O.M. Rosen, Proc. Nat.
Acad. Sci. U.S.A., 68 (1971) 731.

(29) E. Miyamoto, G.L. Petzold, J.S. Harris and P. Greengard, Biochem. Biophys. Res. Comm., 44 (1971) 305.

(30) E. Miyamoto, G.L. Petzold, J.F. Kuo and P. Greengard, J. Biol. Chem., 248 (1973) 179.

(31) M. Tao, Biochem. Biophys. Res. Comm., 46 (1972) 56.

(32) T.P. Lee, J.F. Kuo and P. Greengard, Proc. Nat. Acad. Sci. U.S.A., 69 (1972) 3287.

(33) R.M. Eccles and B. Libet, J. Physiol. Lond., 157 (1961) 484.

(34) J.W. Kebabian, A. Steiner and P. Greengard, manuscript in preparation.

(35) J.F. Kuo and P. Greengard, J. Biol. Chem., 245 (1970) 2493.

(36) P. Greengard and J.F. Kuo, in: Advances in Biochemical Psychopharmacology, Vol. 3, eds. E. Costa and P. Greengard (Raven Press, New York, 1970) p. 287.

(37) J.F. Kuo, G.R. Wyatt and P. Greengard, J. Biol. Chem., 246 (1971) 7159.

(38) F. Hofmann and G. Sold, Biochem. Biophys. Res. Comm., 49 (1972) 1100.

(39) B.J. Hoffer, G.R. Siggins and F.E. Bloom, in: Advances in Biochemical Psychopharmacology, Vol. 3, eds. E. Costa and P. Greengard (Raven Press, New York, 1970) p. 349.

(40) B.J. Hoffer, G.R. Siggins, A.P. Oliver and F.E. Bloom, in: Advances in Cyclic Nucleotide Research, Vol. 1, eds. P. Greengard, R. Paoletti and G.A. Robison (Raven Press, New York, 1972) p. 411.

(41) J.F. Kuo and P. Greengard, J. Biol. Chem., 244 (1969) 3417.

(42) J.F. Kuo and P. Greengard, Proc. Nat. Acad. Sci. U.S.A., 64 (1969) 1349.

ACKNOWLEDGMENT

The original work described in this article was supported by United States Public Health Service Grants NS 08440
and MH 17387.

Figure 4 and 5 are reproduced by permission of Science,
figure 6 by permission of Raven Press, N.Y., and figures
7 - 12 by permission of the Journal of Biological Chemistry.

DISCUSSION

M. BITENSKY: Do you feel that there are any difficulties
associated with a particulate kinase interacting with a
particulate membrane protein? Or do you assume that one
of the components is particulate and the other is soluble?

P. GREENGARD: This is an important question. We do not
know whether the phosphorylation of the synaptic membrane
protein in fact occurs through a protein kinase in the
membrane or a protein kinase in the cell sap and we were
initially reluctant to make a guess about this. However,
as I mentioned in my formal presentation, we have been able
to show a rapid rate of interaction of endogenous membrane
protein kinase and membrane substrate protein. These
results are quite gratifying because the changes in membrane
potential whose molecular basis we are trying to explain
occur fairly rapidly. These membrane potential changes
occur over a period of a few hundred milliseconds. So it is
nice to have the kinase and the substrate rather close to
each other for that kind of rapid response.

S.H. APPEL: Is there a membrane associated cyclic GMP
protein kinase similar to the membrane associated cyclic
AMP protein kinase that you demonstrated in the synapses?

P. GREENGARD: We do not yet have an answer to that question.
We have found cyclic GMP-dependent protein kinase in mammalian brain and observed that cerebellum was a much better
source of this enzyme than was cerebral cortex (Greengard
and Kuo, in: Role of Cyclic AMP in Cell Function; Vol. 3

of E. Costa and P. Greengard, eds., Advances in Biochemical
Psychopharmacology, Raven Press, New York, 1970; p. 287).
However, it has been difficult to work with the cyclic GMP-
dependent protein kinase activity in mammalian brain, and
therefore we have not carried out the subcellular distribution studies. Within the last few weeks, Hoffman and Sold
(Biochem. Biophys. Res. Comm., 49, 1100; 1972) have confirmed our report that there is a cyclic GMP-dependent protein kinase in cerebellum. Moreover, they were able to find
conditions for increasing the activity of cyclic GMP-dependent protein kinase relative to the activity of cyclic AMP-
dependent protein kinase. Their report may make it much
easier to study cyclic GMP-dependent protein kinases in
mammalian nervous tissue.

S.H. APPEL: Most of us who have been separating synapto-
somes for a period of time know (a) how difficult it is
to have them clean and (b) that they are predominantly a
population of pre-synaptic terminals with very little postsynaptic processes. Your data deals predominantly with
post-synaptic processes and I wondered if you had any other
way of assessing how much of your population of synapto-
somes in which you are looking for membrane bound protein
kinases is really post-synaptic as opposed to pre-synaptic.

P. GREENGARD: That is a point of some concern to us. For
that reason, we have recently done a small collaborative
study with Dr. Steven Kornguth. He has sent us some fractions which are relatively more enriched in post-synaptic
membranes than the preparations which we have been using,
and we were able to get the same results with two types
of preparation.

S.H. APPEL: My last question will provide you with an opportunity to speculate. You postulate that dopamine causes
protein phosphorylation, thereby, giving rise to hyper-po-
larization. On the other hand the cholinergic system also
causes protein phosphorylation (through cyclic GMP) but
causes depolarization. How can protein phosphorylation
cause such opposite effects?

P. GREENGARD: Using model substrate proteins, Dr. J.F. Kuo
and I found (Biochim. Biophys. Acta, 212, 434; 1970) that
the substrate specificities of cyclic GMP-dependent protein

kinase and cyclic AMP-dependent protein kinase from lobster
tail muscle were quite different from one another. Therefore, I would speculate that in mammalian brain the endogenous substrate for cyclic GMP-dependent protein kinase is
different from that for cyclic AMP-dependent protein kinase.
We are looking intensively at the present time for substrate
proteins for the cyclic GMP-dependent protein kinase. It
will be of great interest to know whether or not synaptic
membranes do contain a natural substrate protein for cyclic
GMP-dependent protein kinase.

G.A. KIMMICH: Are you aware of any relationship between
the phosphorylated components which you observe and those
that have been described and thought to play a role in the
function of the electrogenic sodium pump?

P. GREENGARD: We have considered the possibility that ATP-
ase in the synaptic membrane might serve as substrate for
cyclic AMP-dependent protein kinase. Preliminary experiments suggest that this is probably not the case.

G.A. KIMMICH: You have not investigated whether monovalent
ions effect the degree of phosphorylation in your system?

P. GREENGARD: We have been investigating the effects of
various monovalent cations on the phosphorylation of synaptic membrane proteins, but the results are too complex to
warrant comment at this point.

H.R. MAHLER: I was wondering whether you could give us any
additional information as to the nature of the endogenous
substrate, such details as molecular weight or any other
physical or chemical properties.

P. GREENGARD: The endogenous substrate for the cyclic AMP-
dependent protein kinase in the synaptic membrane fraction
which I discussed in my presentation had an apparent molecular weight, on SDS gels, of slightly less than 100,000.

E.G. KREBS: It seems to me that we have an almost insur-
mountable problem that we are going to encounter time and
again and that is to determine the function of proteins
which are found to be phosphorylateable in cells. This
is difficult enough for soluble proteins, and is going to

173

be extremely difficult with membrane-bound proteins. Would you care to comment on this?

P. GREENGARD: The problem of how to proceed in attempting to establish physiological correlations with phosphorylation of membrane proteins is of considerable concern to us. One approach which we are using is to try to find biological preparations which have time constants for changes in membrane potential and ion transport which are sufficiently long so that we can attempt to correlate alterations in the state of phosphorylation of membrane protein with alterations in membrane potential or ion transport. The superior cervical ganglion is not a suitable preparation for this purpose because of the fairly rapid time constants of the rise and fall of the inhibitory postsynaptic potential.

STUDIES OF THE cAMP RECEPTOR PROTEIN AND OF THE cAMP DEPENDENT PHOSPHORYLATION OF RIBOSOMAL PROTEIN

G. N. GILL, G. M. WALTON, K. E. HOLDY, C. N. MARIASH
and J. B. KALSTROM
Department of Medicine, Division of Endocrinology
University of California, San Diego, School of Medicine
La Jolla, California 92037

Abstract: The cAMP receptor protein functions as the regulatory sub-unit of cAMP dependent protein phosphokinase. In activating the kinase, cAMP binds to the receptor and causes it to dissociate from its complex with the kinase. The kinase, freed of receptor, is fully activated and no longer stimulable by cAMP. The inter-action of cAMP with the receptor is temperature dependent and mildly influenced by pH but not affected by changes in ionic strength. At $23°$, using purified receptor, the association rate constant is $\sim 10^6$ M^{-1} sec^{-1} and the first order dissociation rate constant is $\sim 3 \times 10^{-3}$ sec^{-1}. Ribosomal protein serves as one substrate for cAMP dependent protein kinase. Ribosomes and ribosomal subunits are phosphorylated in a reaction stimulated by cAMP with an apparent K_m of 4×10^{-8} M. Quantitatively, 10 moles of phosphate are incorporated per mole of 60S subunit and 2 moles of phosphate are incorporated per 40S subunit. The phos-phorylated ribosomal proteins from each subunit have been identi-fied on polyacrylamide gel electrophoresis. These events are postulated to be central to hormonal regulation of growth and dif-ferentiated function.

INTRODUCTION

Adenosine 3',5'-monophosphate (cAMP) serves as the intracellular mediator of adrenocorticotropic hormone (ACTH) in the
adrenal cortex, controlling both the differentiated function of
steroidogenesis and the growth and replicative potential of the gland.
Stimulation of steroid hormone biosynthesis by ACTH and cAMP requires protein synthesis (1-6). Inhibition of protein synthesis by
cycloheximide and puromycin blocks or rapidly terminates the stimulation of steroid hormone biosynthesis induced by ACTH and cAMP,
while inhibition of RNA synthesis by actinomycin D is without effect.
cAMP stimulation of the synthesis of a protein involved in the rate-
limiting step of steroid hormone biosynthesis has been inferred from
these studies (2,7). Regulation of protein synthesis by cAMP at a
post-transcriptional level has been implicated also in parotid,
pituitary, and liver tissues (8-10). In normal adrenal cortical tissue
ACTH stimulates growth and replication. After 16 hours of ACTH
administration the incorporation of [^{3}H]thymidine into DNA is
increased and DNA polymerase and thymidine kinase enzymatic
activities are increased (11). DNA content is measurably increased
by 48 hours of ACTH administration. In functional adrenal tumor
cells in tissue culture, ACTH stimulates steroidogenesis in a manner
similar to that observed in normal cells, but inhibits growth and
DNA synthesis. [^{3}H]thymidine incorporation into DNA is severely
inhibited as the cells regain contact-inhibited growth patterns with
altered cellular morphology (12). ACTH, and presumably cAMP,
thus stimulate steroidogenesis in both normal and neoplastic cells
but have differing effects on growth and DNA synthesis in normal
and neoplastic tissue. Thus, the adrenal cortex has provided an
important model for the study of cAMP regulation of differentiated
function and of growth.

FUNCTION OF THE cAMP RECEPTOR PROTEIN

The initial molecular event, after generation of cAMP, is
binding of the nucleotide to a specific receptor protein (13). The
cAMP receptor protein functions as the regulatory subunit of cAMP
dependent protein phosphokinase (E.C.2.7.1.37) (14). The tight
and specific association of cAMP with the receptor protein results

in dissociation of the cAMP bound receptor from the receptor : protein kinase complex; the free kinase which does not bind cAMP is the fully activated form (15-21). This may be expressed as:

$$R : K + cAMP \rightleftharpoons K + cAMP - R,$$

$$(inactive) \qquad (active)$$

where R : K is the inactive receptor : kinase complex, K the active form of protein phosphokinase and cAMP - R the cAMP bound receptor. Study of the purified cAMP dependent protein kinase from adrenal cortical tissue in the analytical ultracentrifuge revealed a 152,000 molecular weight protein containing both cAMP receptor and cAMP-dependent protein kinase activities. Incubation with cAMP resulted in dissociation of the complex into cAMP receptor and cAMP independent protein kinase (Fig. 1). As shown in the upper panel, both cAMP receptor and protein kinase are present in a molecular complex. Incubation with saturating concentrations of cAMP prior to electrophoresis in polyacrylamide gels results in dissociation of the complex into cAMP receptor and protein kinase subunits as shown in the second panel. Molecular weight estimates of each subunit, obtained by electrophoresis at multiple gel con-centrations, revealed a 60,500 molecular weight kinase subunit and a 92,000 molecular weight receptor subunit (19). The kinase subunit no longer binds cAMP and is fully active in the absence of cAMP. In the experiment shown in Fig. 2, equal amounts of protein were electrophoresed in polyacrylamide gels without and with prior incubation with cAMP. The kinase present in the complex, shown in the upper panel of Fig. 1and on the left in Fig. 2, is stimulated by cAMP at each time point, while the isolated kinase subunit shown in the second panel of Fig. 1 and on the right in Fig. 2, is fully active and no longer stimulable by cAMP. Addition of receptor to isolated kinase subunits suppresses activity and restores cAMP responsiveness (15,19,21-23).

The kinetics of the cAMP : receptor protein interaction have been measured utilizing cellulose ester filters to isolate the cAMP receptor complex (24). The equilibrium dissociation constant, K_D, equals approximately 10^{-8} M (13,24). The association rate constant k_a determined at several cAMP and receptor site concentrations is of the order of 10^6 M^{-1} sec^{-1} (Fig. 3). For these experiments, homo-geneous receptor protein free of all kinase activity was utilized.

177

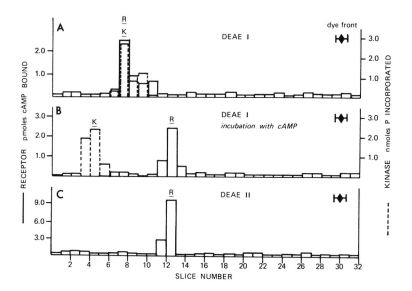

Fig. 1. Effect of cAMP on the receptor: kinase complex.
After electrophoresis in 6% polyacrylamide gels, 2 mm slices
were assayed for kinase and receptor activity. All R_f values
have been corrected for a dye front in slice number 30. Receptor
activity is indicated by the solid line, kinase by the broken line.
A. cAMP dependent protein kinase from DEAE cellulose
chromatography peak I, 68 μg. Receptor and kinase activities
migrate in a single band at multiple gel concentrations. The
separately determined retardation coefficient (K_r) for receptor
activity equals 0.13472 + .012 and for kinase activity equals
0.13761 + .009 indicating identity.
B. cAMP-dependent protein kinase as in A, incubated for 30
minutes on ice with 5×10^{-7} M [^{3}H]cAMP prior to electrophoretic separation. Wide separation of receptor and kinase
activity on the basis of both size and charge has occurred.
C. Free receptor protein derived from DEAE peak II, 23 μg.
The free receptor migrates identically to that obtained from the

receptor : kinase complex. Adapted from Gill, G. N. and Garren, L. D. (19).

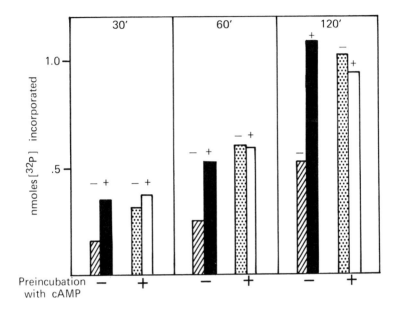

Fig. 2. Effect of cAMP on protein kinase after receptor removal. cAMP-dependent protein kinase was subjected to electrophoresis in 6% polyacrylamide gels as described in Fig. 1. Gel slices corresponding to the R_f of the receptor : kinase complex (Fig. 1A) and to the R_f of the free kinase (Fig. 1B) were assayed for protein kinase activity in the presence and the absence of added cAMP. The receptor : kinase complex is shown on the left and the free kinase is shown on the right at each time point.

    ◨ No cAMP in assay.

    ■ 10^{-6} M cAMP added to assay.

    ▦ Incubation with cAMP prior to electrophoresis, no cAMP in assay.

☐ Incubation with cAMP prior to electrophoresis, 10^{-6} M
　cAMP added to assay.

Addition of cAMP prior to electrophoresis is indicated below the
columns and addition of cAMP to the assay is indicated above
each column. cAMP receptor activity is associated with the
complex, but is absent from the free kinase area. Free cAMP
migrates with the dye front and is absent from the free kinase
area of the gel.

　The left side of each panel demonstrates that cAMP
stimulability can be readily demonstrated in the complex in the
gel. Removal of receptor, shown on the right of each panel,
results in a fully activated protein kinase that is no longer
stimulated by cAMP. Adapted from Gill, G. N. and Garren,
L. D. (19).

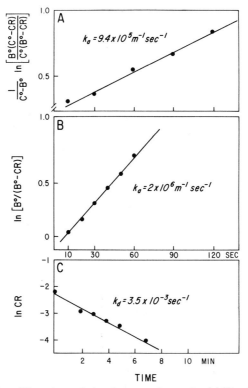

Fig. 3. Kinetics of the interaction of cAMP with receptor
protein. Bovine adrenal cortical cAMP receptor protein was

purified from DEAE cellulose chromatography peak II (15,19) by sucrose gradient centrifugation and polyacrylamide gel electrophoresis. The receptor protein was removed from acrylamide by electrophoresis into agarose. No kinase activity was detectable in the final receptor protein preparation.

A. Linear plot of the association kinetic data. The data shown are plotted in the integrated form of a second order reaction:

$$k_a t = \frac{1}{C^o - B^o} \ln \left[\frac{B^o (C^o - CR)}{C^o (B^o - CR)} \right]$$

where C^o and B^o represent the initial concentrations of cAMP and of binding sites, respectively, and CR represents the concentration of bound complex at any time t. The maximum number of binding sites, B^o, at t = 0, was determined as the amount of cAMP bound under saturating conditions. The experiments shown were performed at 23° in standard buffer with $B^o = 1.38 \times 10^{-9}$ M and $C^o = 1.03 \times 10^{-9}$ M. Similar results were obtained using several cAMP and receptor site concentrations. As far as tested, all of the association data are consistent with a reaction which is first order with respect to both cAMP and receptor sites.

B. Association kinetic data obtained under pseudo first order reaction conditions. Pseudo first order reaction conditions were utilized with C^o 10-30-fold greater than B^o. The data are plotted in the form of a pseudo first order reaction:

$$k_a t = \frac{1}{C^o} \ln \left[\frac{B^o}{B^o - CR} \right]$$

The observed k_a resembles that obtained using the full second order rate equation.

C. Linear plot of dissociation data. The results are plotted according to the rate equation for first order dissociation: $k_d t = \ln CR^o - \ln CR$, where k_d = the first order dissociation rate constant, CR^o = the concentration of bound cAMP at t = 0 and was determined by extrapolating the best linear fit of the data to t = 0. Similar plots were obtained under all reaction conditions and also when gel filtration techniques were used to separate bound from free cAMP. From Gill, G. N., Kalstrom, J. B., Holdy, K. E. and Mariash, C. N. (25).

Association kinetic data obtained under second order or pseudo first order reaction conditions yielded similar results (Fig. 3A, B). The rate of association increases with increasing temperature and increases with lower pH, but is not affected by variations in ionic strength.

The rate of dissociation was determined by allowing the standard receptor to reach equilibrium with [^{3}H]cAMP at least half saturating the receptor; at t = 0, 1,000-fold or more excess unlabeled cAMP was added; after further incubation for varying time periods, the dissociation reaction was stopped by isolating the cAMP receptor complex on cellulose ester filters. The initial dissociation of the cAMP : receptor complex is first order (Fig. 3C). The dissociation reaction is highly temperature dependent being 100-fold faster at 37° than at 0°. Over the ranges tested, neither variation in ionic strength nor changes in pH alter the rate of dissociation. At 37° the cAMP : receptor complex dissociates with a half life of approximately 2.5 minutes. Thus, at physiological temperatures, the dissociation is rapid enough to allow a dynamic equilibrium to be maintained such that alterations in the concentration of free cAMP will alter the amount of bound cAMP.

From the change of the rate constants and the equilibrium constant with temperature, certain energy and thermodynamic parameters of the cAMP : receptor interaction can be calculated. Complete experiments similar to those presented in Fig. 3 but utilizing cAMP receptor at an earlier stage of purification were performed at several temperatures and linear Arrhenius and van't Hoff plots derived (26). The thermodynamic data for the cAMP : receptor interaction are presented in Table 1. The Arrhenius activation energy is 15 kcal mole^{-1} for association and is 22 kcal mole^{-1} for dissociation. From the change in the equilibrium constant, k_d/k_a, with temperature, the enthalpy change for binding, Δ H, is determined to be 9.2 kcal mole^{-1}. The free energy change, Δ G, for the reaction can be calculated at any temperature from the measured equilibrium constant. At 25° C, Δ G is - 11 kcal mole^{-1}. The entropy change, Δ S, is then calculated to be 68 cal mole^{-1} °K^{-1}. The increase in entropy is thus the driving force for the reaction.

182

TABLE 1

Energy parameters of the cAMP-receptor interaction

Activation energy[a]	
Association	15 kcal/mole
Dissociation	22 kcal/mole
Thermodynamic parameters (at 25° C)	
Δ G[b]	- 11 kcal/mole
Δ H[c]	9.2 kcal/mole
Δ S[d]	68 cal mole^{-1} °K^{-1}

[a] Calculated using the Arrhenius equation (26).
[b] Δ G = RTln $(1/K_d)$.
[c] Calculated from the van't Hoff equation (26).
[d] Δ S = (Δ H - Δ G)/T.

cAMP-DEPENDENT PHOSPHORYLATION OF RIBOSOMAL PROTEIN

A number of substrates for cAMP-dependent protein kinase have been identified. cAMP-dependent protein kinase catalyzed phosphorylation of phosphorylase kinase and of glycogen synthetase results in marked alteration of enzyme activity with activation of the first and inactivation of the second enzyme (27). Similar cAMP-dependent phosphorylation of hormone sensitive lipase results in activation of that enzyme (28,29). Phosphorylation of histone, protamine, microtubules, and synaptic membranes also occurs (30-33). The cAMP-dependent protein kinase from adrenal cortical cells is localized with highest specific activity in the endoplasmic reticulum (34). Because phosphorylation of a substrate present in microsomes had been observed (34,35), ribosomes were tested as substrate for the isolated enzyme. Adrenal cortical ribosomes prepared by the method of Takanani (36) were incubated with cAMP-dependent protein kinase purified from the cytosol (Fig. 4). Extensive cAMP-dependent phosphorylation of ribosomal protein

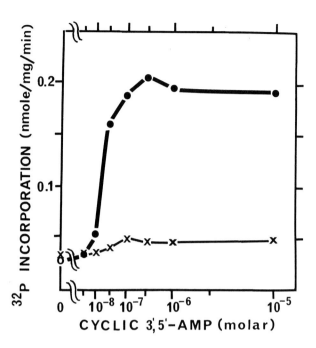

Fig. 4. Effect of cAMP on protein kinase catalyzed phosphorylation of ribosomal protein. Adrenal cortical ribosomes, 0.2 mg, were incubated with cAMP-dependent protein kinase, 40 µg, in standard protein kinase assay mixtures containing 50 mM glycerol phosphate pH 6.0, 20 mM NaF, 4 mM theophylline, 10 mM $MgCl_2$, 0.6 mM γ [^{32}P]ATP and increasing concentrations of cAMP. Incorporation of ^{32}P into hot TCA insoluble protein was measured after incubation for 10 min. at 30°. Endogenous phosphorylation of ribosomes in the absence of added enzyme has been subtracted.

●——● Kinase - catalyzed ^{32}P incorporation into ribosomal protein.

✗——✗ Endogenous phosphorylation of the enzyme preparation.

Apparent K_m for cAMP = 4 X 10^{-8} M. From Walton, G. M., Gill, G. N., Abrass, I. B., and Garren, L. D. (34).

occurs with an apparent K_m for cAMP of approximately 4×10^{-8} M. The ^{32}P is incorporated into hot TCA insoluble material; high voltage electrophoresis of hydrolyzed ribosomal protein indicated phosphorylation of serine and threonine residues (34). cAMP and protein kinase dependent phosphorylation of protein associated with each of the ribosomal subunits occurs (Fig. 5).

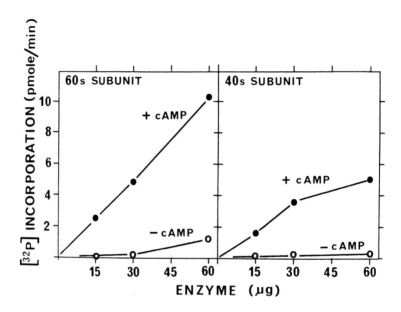

Fig. 5. Phosphorylation of isolated 60S and 40S ribosomal subunits as a function of added protein kinase. 2.6 A_{260} units of 60S or 2.8 A_{260} units of 40S ribosomal subunits were incubated in the standard kinase assay with increasing concentrations of cAMP-dependent protein kinase. Incubations were for 10 min. at 30° in the presence and absence of cAMP. Incorporation into hot TCA insoluble protein was determined. Values are corrected for endogenous incorporation into enzyme and subunits in the presence and absence of cAMP. From Walton, G. M. and Gill, G. N. (37).

Because ribosomes as isolated contain phosphoproteins, full quantitation of the extent of ribosomal protein phosphorylation is uncertain. As estimation of the extent of ribosomal protein phosphorylation possible in the in vitro reaction was obtained by complete phosphorylation of each of the isolated subunits (Table 2). Quantitatively the 60S subunit contains 10 moles and the 40S subunit 2 moles of ^{32}P per mole of subunit. Specific ribosomal protein bands were phosphorylated on each of the subunits (Fig. 6).

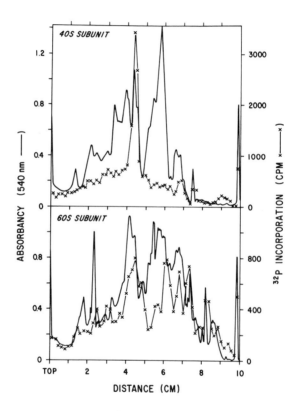

Fig. 6. Polyacrylamide gel electrophoresis of protein extracted from 60S and 40S subunits. Phosphorylation was performed for 80 min. in a volume of 2.5 ml containing 50 mM

TABLE 2

Complete phosphorylation of 60S subunit and 40S subunit in the protein kinase catalyzed reaction [a]

Substrate	Protein content (mg/A_{260} unit)	Absorbance ratio A_{260}/235	32P Incorporation [b] (nmoles/A_{260} unit)	moles 32P/ mole of subunit
60S subunit	0.030 ± 0.002 (3)	1.67 ± 0.03 (3)	0.213 ± 0.053 (6)	10
40S subunit	0.051 ± 0.005 (3)	1.61 ± 0.02 (3)	0.158 ± 0.016 (6)	2

[a] Phosphorylation was performed in reaction mixtures containing glycerol phosphate (pH 6.0), 2 mM $MgCl_2$ and 6.8-11 .6 μg of enzyme (SG peak III). Specific activity of [32P]ATP was 2.6 - 8.5 X 10^4 cpm/nmole. Amounts of substrate varied between 0.8 and 3.9 A_{260} units. Incubation time was 2 hrs. Values are means ± standard deviations with number of determination in parenthesis, and incorporation corrected for 0.015 nmoles of endogenous incorporation per 10 μg of enzyme.

[b] Molecular weights of 3.0 X 10^6 and 1.5 X 10^6 and protein content of 43 and 55% for the 60S and 40S subunits respectively were used for molar incorporation calculations (38). From Walton, G. M. and Gill, G. N. (37).

Tris-HCl (pH 7.5), 1 mM dithiothreitol, 2 mM $MgCl_2$, 5 μM cAMP, 0.3 mM [^{32}P]ATP (9.62 X 10^4 cpm/nmole), 450 μg of DEAE peak III enzyme and 52 A_{260} units of 40S subunit or 49 A_{260} units of 60S subunit. Reactions were stopped with an equal volume of 14 mM ATP (pH 7.5). Ribosomal subunits were isolated from reaction mixtures by discontinuous sucrose gradient sedimentation; protein was extracted with 6 M LiCl and 8 M urea (39). 260 μg and 300 μg of protein from 60S and 40S subunits respectively were used for polyacrylamide gel electrophoresis in a buffer system containing 8 M urea and β-alanine-acetic acid (pH 4.5). From Walton, G. M. and Gill, G. N. (37).

Approximately 8 proteins on the 60S subunit and a single band on the 40S subunit are phosphorylated in the cAMP and protein kinase dependent reaction.

In all mammalian systems studied, ribosomal phosphoproteins have been identified (40-42). Kabat has indicated that the phosphate groups of reticulocyte ribosomes turn over at a rate of approximately 3% per minute (43). The phosphate groups are present on a limited number of ribosomal proteins (40-44) and the hormonal state appears to influence the phosphorylation of ribosomal protein. Treatment of rats with glucagon resulted in increased phosphorylation of ribosomal proteins with one protein band being specifically increased 2-3-fold (45). Thyroidectomy resulted in a 35% decrease in the phosphate content of ribosomal proteins; 3,5,-3 triiodothyronine administration restored the phosphate content to control levels (46).

The functional role of ribosomal protein phosphorylation has not been defined. Phosphorylation of ribosomal protein may modify the structure of the ribosome by altering the protein : nucleic acid interaction in a manner analogous to phosphorylation of histone which results in modification of the histone : DNA interaction (47,48). This modification may be involved in ribosomal assembly or in ribosome function. Phosphorylation of specific ribosomal proteins may change the functional state of the protein as has been demonstrated for the enzymes involved in glycogen metabolism. cAMP-dependent phosphorylation of ribosomal protein is thus one potential mechanism for hormonal control of protein synthesis at a translational level.

COMMENTS

The initial events in cAMP action, interaction with the specific receptor and control of cAMP-dependent protein kinase, appear common to all eukaryotic systems studied. The cAMP receptor and cAMP-dependent protein kinase present in functional adrenal tumor cells in tissue culture resemble those found in normal adrenal cells. It is likely that this initial mechanism results in control of steroidogenesis in both normal and neoplastic adrenal cortical cells and it is also likely that this common initial mechanism results in widely differing effects on growth with stimulation in normal and inhibition in neoplastic cells. Identification of events regulated by the cAMP-dependent protein kinase and cAMP receptor should provide insight into control of the expression of differentiated function and of growth. The cAMP-dependent phosphorylation of ribosomal protein may be involved in each of these processes.

REFERENCES

(1) J. J. Ferguson, Jr., J. Biol. Chem. 238 (1963) 2754.

(2) L. D. Garren, R. L. Ney and W. W. Davis, Proc. Nat. Acad. Sci. USA 53 (1965) 1443.

(3) R. V. Farese, Biochem. Biophys. Acta 87 (1964) 699.

(4) G. H. Sato, T. Rossman, L. Edelstein, S. Holmes and V. Buonassisi, Science 148 (1965) 1733.

(5) J. Kowal, Endocrinology 87 (1970) 951.

(6) D. Schulster, S. A. S. Tait, J. F. Tait and J. Mrotek, Endocrinology 86 (1970) 487.

(7) L. D. Garren, G. N. Gill, H. Masui and G. M. Walton, Recent Prog. Hormone Res. 27 (1971) 433.

(8) R. J. Grand and P. R. Gross, Proc. Nat. Acad. Sci. USA 65 (1970) 1081.

(9) F. Labrie, G. Beraud, M. Gauthier and A. Lemay, J. Biol. Chem. 246 (1971) 1902.

(10) W. D. Wicks, J. Biol. Chem. 246 (1971) 217.

(11) H. Masui and L. D. Garren, J. Biol. Chem. 245 (1970) 2627.

(12) H. Masui and L. D. Garren, Proc. Nat. Acad. Sci. USA 68 (1971) 3206.

(13) G. N. Gill and L. D. Garren, Proc. Nat. Acad. Sci. USA 63 (1969) 512.

(14) D. A. Walsh, J. P. Perkins and E. G. Krebs, J. Biol. Chem. 243 (1968) 3763.

(15) G. N. Gill and L. D. Garren, Biochem. Biophys. Res. Comm. 39 (1970) 335.

(16) M. Tao, M. L. Salas and F. Lipmann, Proc. Nat. Acad. Sci. USA 67 (1970) 408.

(17) A. H. Kumon, D. Yamamura and Y. Nishizuka, Biochem. Biophys. Res. Comm. 41 (1970) 1290.

(18) E. M. Reimann, C. O. Brostrom, J. D. Corbin, C. A. King and E. G. Krebs, Biochem. Biophys. Res. Comm. 42 (1971) 187.

(19) G. N. Gill and L. D. Garren, Proc. Nat. Acad. Sci. USA 68 (1971) 786.

(20) J. Erlichman, A. H. Hirsch and O. M. Rosen, Proc. Nat. Acad. Sci. USA 68 (1971) 731.

(21) L. D. Garren, G. N. Gill and G. M. Walton, Ann. N. Y. Acad. Sci. 185 (1971) 210.

(22) H. Yamamura, A. Kumon and Y. Nishizuka, J. Biol. Chem. 246 (1971) 1544.

(23) C. O. Brostrom, J. D. Corbin, C. A. King and E. G. Krebs, Proc. Nat. Acad. Sci. USA 68 (1971) 2444.

(24) G. M. Walton and L. D. Garren, Biochemistry 9 (1970) 4223.

(25) G. N. Gill, J. B. Kalstrom, K. E. Holdy and C. N. Mariash, manuscript in preparation.

(26) W. J. Moore, Physical Chemistry, Third Edition (Prentice-Hall, Inc., Englewood Cliffs, N. J., 1962).

(27) T. R. Soderling, J. P. Hickenbottom, E. M. Reimann, F. L. Hunkeler, D. A. Walsh and E. G. Krebs, J. Biol. Chem. 245 (1970) 6317.

(28) J. K. Huttunen, D. Steinberg and S. E. Mayer, Proc. Nat. Acad. Sci. USA 67 (1970) 290.

(29) J. W. Corbin, C. O. Brostrom, R. L. Alexander and E. G. Krebs, J. Biol. Chem. 247 (1972) 3736.

(30) T. A. Langan, J. Biol. Chem. 244 (1969) 5763.

(31) B. Jergil and G. H. Dixon, J. Biol. Chem. 245 (1970) 425.

(32) D. B. P. Goodman, H. Rasmussen, F. DiBella and C. E. Guthrow, Jr., Proc. Nat. Acad. Sci. USA 67 (1970) 652.

(33) E. M. Johnson, H. Maeno and P. Greengard, J. Biol. Chem. 246 (1971) 7731.

(34) G. M. Walton, G. N. Gill, I. B. Abrass and L. D. Garren, Proc. Nat. Acad. Sci. USA 68 (1971) 880.

(35) M. Weller and R. Rodnight, Nature 225 (1970) 187.

(36) M. Takanami , Biochem. Biophys. Acta 39 (1960) 318.

(37) G. M. Walton and G. N. Gill, submitted for publication.

(38) M. G. Hamilton, A. Pavlovec and M. J. Petermann, Biochemistry 10 (1971) 3424.

(39) P. S. Leboy, E. C. Cox and J. G. Flaks, Proc. Nat. Acad. Sci. USA 52 (1964) 1367.

(40) D. Kabat, Biochemistry 9 (1970) 4160.

(41) J. E. Loeb and C. Blat, Fed. Eur. Biochem. Soc. Letters 10 (1970) 105.

(42) C. Eil and I. G. Wool, Biochem. Biophys. Res. Comm. 43 (1971) 1001.

(43) D. Kabat, J. Biol. Chem. 247 (1972) 5338.

(44) L. Bitte and D. Kabat, J. Biol. Chem. 247 (1972) 5345.

(45) C. Blat and J. E. Loeb, Fed. Eur. Biochem. Soc. Letters 18 (1971) 124.

(46) G. Correze, P. Pinell and J. Nunez, Fed. Eur. Biochem. Soc. Letters 23 (1972) 87.

(47) T. A. Langan, in: Regulatory Mechanisms for Protein Synthesis in Mammalian Cells, eds. A. San Pietro, M. R. Lamborg and F. T. Kenny (Academic Press, New York, 1968) pp. 101–118.

(48) M. T. Sung and G. H. Dixon, Proc. Nat. Acad. Sci. USA 67 (1970) 1616.

This investigation was supported in part by PHS Research Grants No. AM13149 and No. AM14415 from the Institute of Arthritis and

Metabolic Diseases. GNG is a recipient of PHS Research Career Development Award No. AM70215 from the Institute of Arthritis and Metabolic Diseases.

Fig. 1,2 and 4 are reproduced by permission of the National Academy of Sciences.

DISCUSSION

H. ROSENKRANTZ: The scheme that you drew on the blackboard seems to fit the facts that you have dealt with at the moment. There are two observations that have been reported over the last fifteen years which have not been handled at all, and I was hoping that you would make some comment on them. Firstly, there have been a number of substances that are not ACTH derivatives that have been shown to increase steroidogenesis in the adrenal cortex. I mention two such substances, serotonin, and angiotonin; there are others that have been reported in the literature. Secondly, the hypothesis that you illustrate on the blackboard indicates that ACTH must have its function through membrane bound adenyl cyclase. ACTH has also been shown to work on beef adrenal homogenates in stimulating the conversion of cholesterol to pregnenolene. This work was reported many years ago by workers at the Worcester Foundation. I wonder how you can combine these two ancillary observations on hormone stimulation in the adrenal cortex with the present work going on with cyclic AMP.

G.N. GILL: Do you know whether the first two substances increase cyclic AMP in the adrenal cortex?

H. ROSENKRANTZ: No, I do not believe that has ever been studied. The only thing that was measured was the increase in the production of corticosterone. This was done with quartered rat adrenals, and it has been confirmed by adrenal vein cannulation in the dog where serotonin has been shown to increase compound F, cortisol.

G.N. GILL: It would be interesting to see what cyclic AMP levels were in response to these because prostoglandin

E_1, for instance, will stimulate steriodogenesis, but this appears to be through exactly the same cyclic AMP mechanism. In regard to your second question, I really cannot comment on that. Most workers have not been able to reproduce any hormonal effects in homogenates or in cell-free preparation with any of the polypeptide hormones reproducibly.

B.L. BROWN: In relation to the earlier question by Dr. Rosenkrantz, we have recently shown, in collaboration with J.F and S.A.S. Tait, that serotonin and increased potassium concentration (3.6 mM to 8.4 mM) do increase the cyclic AMP in incubations of isolated zona glomerulosa cells, but not in fasciculata cells. (Albano, J., Brown, B.L., Elkins. R.P., Price, I., Tait. S.A.S. and Tait, J.F. Journal of Endocrinology. 1973, in press).

M.S. ROSE: I wonder if you would care to speculate on the role of calcium in you system. I think this has been shown quite explicitly to be involved in eliciting the ACTH response on steroidogenesis in the adrenal cortex.

G.N. GILL: I can only comment on the work of others. Two facts seem well established. Calcium, while not required for ACTH binding to the plasma membrane of the cell, is required for the stimulation of the adenyl cylase. Secondly, cyclic AMP is required at an additional step which appears to be in some way involved with protein synthesis.

A.G. GORNALL: I want to call attention to recent work of a colleague of mine, Dr. C.C. Liew. He has found that, in addition to phosphorylation, there is also rapid and extensive acetylation of ribosomal proteins. When rat liver ribosomes were pulse labeled in vivo with ^{3}H-acetate, maximum incorporation occurred within 15 minutes; monomers and all classes of polyribosomes were labeled. After 0.5 M KCl and puromycin treatment, and centrifugation in a sucrose gradient, the ribosomal subunits were isolated. Specific activity in the small subunit was somewhat higher than that of the large subunit. Approaching the problem in a different way, ribosomal RNA was removed by urea-LiCl treatment and the proteins were then separated on 10% polyacrylamide gel. Acetylation of ribosomal proteins appeared to be greater in certain fractions. Preliminary data suggest that acetylation may play a regulatory role at the translational level.

J.M. MARSH: I would like to suggest that perhaps the phosphorylation of ribosomes might be associated more with the effect of ACTH on the long term growth of the adrenal rather than the accute effect of ACTH on the stimulation of steroidogenesis. You indicated in your abstract that ACTH causes the phosphorylation of ribosomes in vivo as well as in vitro. Have you done a time study where you looked at the effect of phosphorylation of the ribosome in relationship to the increase of steroidogenesis?

G.N. GILL: No, we have not. Dr. Bernard Roos has looked at the in vivo reaction in response to ACTH. He has not been able to do a early time curve predominatly because of pool problems. As you know almost all hormones increase the uptake of P^{32} into cells very rapidly, so you have to wait for a period of equilibrium to take care of the pool effects. I really cannot define the time course of the in vivo phosphorylation, but only state that it certainly occurs.

R. SHARMA: Are other cyclic nucleotides active in your system?

G.N. GILL: As I mentioned, the structural similarity of cyclic nucleotides to cyclic AMP determines their ability to bind to the receptor. Cyclic IMP being the closest to cyclic AMP is obviously the closest to producing the effects of cyclic AMP. Cyclic GMP is about one or two orders of magnitude different in terms of its ability to compete with cyclic AMP for the receptor and generally requires much larger concentrations to reproduce the same in vivo effects.

C. ABELL: Does the phosphorylation of the ribosomal proteins have any influence on reconstitution of ribosomal sub-units?

G.N. GILL: I cannot answer that from our own work. From the work of Dr. Ira Wool, it would appear that it does not.

195

E.G. KREBS: I would like to ask whether or not it would be worth while looking at the calcium dependent protein kinase namely phosphorylase kinase, to see whether or not it is acting in the adrenal?

G.N. GILL: I would like very much to do that.

G. KRISHNA: What do you think is the role of the protein of which the synthesis is blocked by cycloheximide in the control mechanism of the steroidognesis, by ACTH?

G.N. GILL: I think that there are at least three conflicting theories that have in part been brought up here. Dr. Marsh, showed a slide yesterday of possible sites, and I think in all honesty that no one knows what the rate limiting step in steriod hormone production is, either in the adrenal cortex or in the ovary or in the testes. These three tissues are quite alike, in their sub-cellular organization, and in the site of hormonal control of steroidogenesis, that is between the conversion of cholestrol to pregneneolone. I would be glad to speculate, but it would only be that, the rate limiting step involves a cholesterol carrier protein, which carries cholosterol from the cytosol into the mitochondria where the desmolase enzyme system is located.

G. KRISHNA: Do you have any idea whether mitochondrial protein synthesis is involved? Have you tested inhibitors of mitochondrial protein synthesis such as chloramphenicol?

G.N. GILL: These studies were done by Garren a number of years ago, and as a matter of fact, mitochondrial protein synthesis does not appear to be involved in the rate limiting step. It appears that microsomal protein synthesis inhibitors rather than chloramphenicol. block hormonal stimulation of steroidogenesis.

G.H. DIXON: From your discussion, I assume that you are thinking that the phosphorylation of preexisting ribosomes occurs, is there any possiblity that what you might be seeing is newly assembled ribosomes, and that this phosphorylation may have something to do with assembly of ribosomes?

G.N. GILL: I really think we do not know the physiology
of the phosphorylation yet. I think that the association
and disassociation of proteins from the ribosomes by phosphorylation may be involved in the process. Ribosomal
assembly usually takes place before the ribosomes are
transported into the cytoplasm, however, these are cytoplasmic ribosomes that we are looking at.

M. TAO: I seems to me that to study the significance of
phosphorylation on riobsomes one would have to deal with
the dephosphorylation reaction of ribosomes because I would
assume that if you isolate ribosmes from the cells they are
in the phosphorylated form as you have indicated.

G.N. GILL: I think that the extent of phosphorylation that
we calculate represents a minimal figure. We have only
determined the phosphate content of whole ribosomes so far.
We have not determined the endogenous content of the subunits that we are using, but if we extrapolate from the
whole ribosome these have to be minimal figures assuming
that there is some cold phosphate already there. These
patterns of protein phosphorylation however, are quite re-
producible in terms of what is available to the in vitro
cyclic AMP dependent protein kinase reaction.

W.D. WICKS: Have you looked for ribosomal protein phosphorylation in the isolated adrenal tumor cells?

G.N. GILL: Yes, and it does occur, and it does occur in
response to ACTH.

197

PHOSPHORYLATION OF RNA POLYMERASE IN E. COLI AND RAT LIVER

O.J. MARTELO
Department of Medicine
University of Miami School of
Medicine and VA Hospital

Abstract: A cyclic-AMP dependent rabbit muscle protein kinase catalyzes the phosphorylation of Escherichia coli RNA polymerase. The phosphorylation occurs on a serine residue of the σ factor. The stimulation of RNA synthesis and phosphorylation are inhibited by a specific kinase inhibitor. The results suggest that the cyclic AMP and protein kinase may play a role in RNA synthesis via a phosphorylation mechanism. Preliminary studies suggest that a similar mechanism may be operative in rat liver cells.

I. IN E. COLI

INTRODUCTION

RNA synthesis in bacteria is regulated by many different factors. In stringent control the regulation appears to involve two small nucleotides. Recently, Cashel & Gallant (1) have shown that cells undergoing stringent control accumulate ppGpp and the regulation appears to be at the level of RNA polymerase.

Other factors appear to interact with the DNA template. These factors include the lac repressor (2) and the cyclic AMP binding or receptor protein discovered by Zubay & coworkers (3) and Emmer et al. (4).

Another group of regulatory factors influence chain initiation, elongation and termination and include the sigma factor of Burgess (5), the M factor of Davison (6) and the rho factor of Roberts (7).

This work was supported by a Grant-in-Aid (71-881) from the American Heart Association and by a Clinical Investigatorship Grant VA Hospital, Miami, Florida.

Regulation of RNA synthesis may also be influenced by covalent modification of the RNA polymerase molecule. These include inhibition of the E. coli RNA polymerase by adenylation (8) and α subunit adenylation after T4 infection (9). Recent experiments from our laboratory have implicated a phosphorylation-dephosphorylation reaction which may play an important role in the regulation of RNA synthesis in E. coli (10). In the present communication we wish to report in further detail the phosphorylation of E. coli polymerase by a cyclic AMP dependent kinase from rabbit skeletal muscle.

METHODS

Protein kinase from rabbit muscle was prepared by the method of Walsh (11) as modified by Reimann (12). The RNA polymerase from E. coli was prepared by the method of Chamberlin & Berg (13) as modified by Richardson (14). A polyacrylamide-SDS gel electrophoresis pattern for a typical RNA polymerase preparation is shown in Figure 1. Four distinct bands are present and these correspond to the β, β¹, σ and α subunits. Polyacryalamide gel electrophoresis (0.1% SDS) were prepared by the method of Weber & Osborn (15). The assay conditions for RNA synthesis and polymerase phosphorylation were identical to those described previously (10). Sigma factor was obtained as described by Burgess (5).

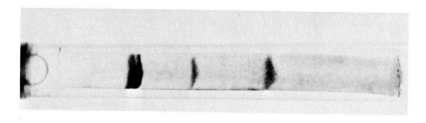

β,β'   σ   α

Figure 1. Polyacrylamide gel electrophoresis of 50 μg of RNA polymerase in SDS.

Repeated assays for polynucleotide phosphorylase activity in the polymerase preparation failed to reveal any contaminating activity.

RESULTS

The stimulation of RNA synthesis by protein kinase is shown in Table 1. A specific inhibitor of rabbit muscle protein kinase inhibits the RNA stimulation by protein kinase.

TABLE 1

Effect of protein kinase on RNA synthesis

Additions	AMP Incorporation (nmoles/10 min)
Blank	0.96
Polymerase	15.2
Polymerase + kinase	40.0
Polymerase + kinase + cyclic AMP	46.0
Polymerase + kinase + cyclic AMP + inhibitor	24.0

The reaction mixture was identical to those described in reference 10 except as follows: 50 µg/ml protein kinase, 50 µg/ml protein kinase inhibitor 2 µm cyclic AMP where indicated.

The phosphorylation of RNA polymerase by protein kinase was also inhibited by the kinase inhibitor as shown in Table 2.

The stoichiometry of ^{32}P per mole of σ factor and core enzyme was calculated. Phosphorylation under these conditions was about 1 mole ^{32}P per 2 moles σ factor and 1 mole of ^{32}P per 40 moles of core enzyme.

Protein kinase can also phosphorylate RNA polymerase utilizing γ^{32}-P-GTP as phosphate donor as shown in Table 3.

TABLE 2

Effect of inhibitor on RNA polymerase phosphorylation

Additions	^{32}P Incorporation (pmole)
Polymerase	2.1
Polymerase + kinase	9.2
Polymerase + kinase + cyclic AMP	59.2
Polymerase + kinase + cyclic AMP + inhibitor	8.4
Sigma	0.1
Sigma + kinase	4.0
Sigma + kinase + cyclic AMP	36.0
Sigma + kinase + cyclic AMP + inhibitor	3.6

The reaction mixture contained: 0.05 M Tris HCl,
pH 7.8, 5 mM $MgCl_2$, 1mM DTT, 0.2 mM γ^{32}-P-ATP,
2 µM cyclic AMP (where indicated), RNA polymerase
1 mg/ml, 80 µg σ factor, 50 µg protein kinase and
50 µg kinase inhibitor.

TABLE 3

Phosphorylation of RNA polymerase by kinase & GTP

Additions	^{32}P Incorporation (pmole)
Polymerase + cyclic AMP	1.7
Polymerase + kinase	3.5
Polymerase + kinase + cyclic AMP	22.0
Sigma + cyclic AMP	0
Sigma + kinase	5.5
Sigma + kinase + cyclic AMP	155

The reaction mixture was identical to that
described in methods except that γ^{32}-P-ATP
was replaced by 0.25 mM γ-^{32}-P-GTP.

Site of Phosphorylation:

Further evidence of the site of phosphorylation is presented in Figure 2. The RNA polymerase was first phosphorylated as described under Table 2 and subjected to gel filtration on G-75 Sephadex to remove unreacted ^{32}P followed by phosphocellulose chromatography to separate σ factor from the core enzyme as described by Burgess (5). The results are shown in Figure 2.

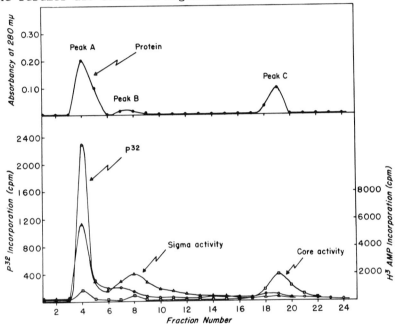

Figure 2. Sigma factor was assayed in the presence of core enzyme using T4DNA or template as described in methods. Core enzyme was assayed with calf thymus DNA as template. Upper panel Peak A: σ factor Peak B:σ + core enzyme. Peak C: core enzyme. Lower panel shows ^{32}P (o—o) or σ activity (Δ—Δ) and core activity (□—□).

It is clear that peak A contains primarily σ factor and nearly all the radioactivity.

Further identification of the phosphorylated site on
RNA polymerase was made by polyacrylamide gel electrophoresis of the ^{32}P-labeled RNA polymerase. In this experiment (Fig. 3) the various subunits of RNA polymerase were
separated by SDS, the gels were stained, sliced longitudi-
nally, embedded in filter paper and subjected to radioauto-
graphy to determine the location of ^{32}P. Protein kinase
20 µg (experiment 1) phosphorylates the σ factor primarily.
At higher concentration of protein kinase 200 µg (experiment 2) phosphorylation is also observed on the core enzyme.

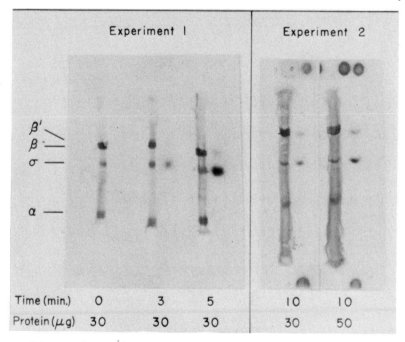

Figure 3. Polyacrylamide gel electrophoresis of
^{32}P-labeled RNA polymerase. SDS gels were sliced,
embedded in Whatman #3 paper and covered with a
Kodak RP x-omatic X-ray film for 24 hours. Film
shifted to right to show the position of the
radioactivity relative to polymerase subunits.

The site of phosphorylation of various acceptors by
protein kinase usually has been the hydroxyl group of
serine or threonine. The phosphate in the 0-phospho-
serine and threonine bond is stable to dilute acid and

labile in dilute alkali (16). This proved to be the case
in acid-base stability studies of the ^{32}P-labeled polymerase (not shown). However, paper electrophoresis of a
partial acid hydrolysate of ^{32}P-labeled RNA polymerase
led to the identification of radioactive phosphoserine
in RNA polymerase as shown in Figure 4.

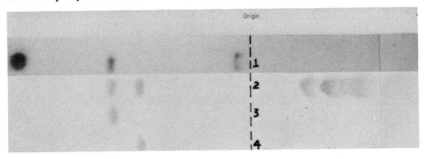

Figure 4. Paper electrophoresis of partial acid
hydrolysate of ^{32}P-labeled polymerase. Phosphorylation of RNA polymerase 1 mg/ml was obtained as
described under Table 2. Position 1 - autoradio-
graph of electrophoresis in Position 2. Position 2
- hydrolysate of ^{32}P-Polymerase plus added phosphothreonine and phosphoserine. Position 3 - phosphoserine standard. Position 4 - Phosphothreonine
standard.

The results demonstrate that phosphorylation of RNA
polymerase by protein kinase results in covalent binding
of phosphate to a serine hydroxyl group in the σ factor.

DISCUSSION

These experiments provide additional evidence that
phosphorylation of the σ factor of E. coli RNA polymerase
correlates with the increased activity of this enzyme with
T4 DNA as template. The fact that the stimulation of RNA
synthesis and phosphorylation of RNA polymerase by protein
kinase are inhibited, by a specific inhibitor of protein
kinase provide strong evidence that the increase in RNA
polymerase activity is due to a phosphorylation.

A study of the effects of cyclic AMP on β-galactosidase biosynthesis has led to the isolation of a cyclic AMP binding protein from E. coli (3,4). It has been reported that this protein has no kinase activity and appears to play a role in RNA regulation by binding to the DNA template.

Cyclic AMP binding proteins have been isolated from mammalian tissues and in these cases the receptor protein is identical to the regulatory subunit of protein kinase (17,18). Thus, the cyclic AMP binding protein from E. coli appears to be quite different from the mammalian kinases which bind cyclic AMP. It is clear that further studies are necessary to clarify the relationship, if any between E. coli cyclic AMP binding protein and cyclic AMP dependent protein kinase.

II. IN RAT LIVER

INTRODUCTION

Regulation of gene expression in mammalian cells is complex. Phosphorylation of histone with subsequent derepression of template activity is an important regulatory mechanism (19). Histone acetylation appears to have similar function (20). In addition, a phosphorylation of certain nuclear (non-histone) acidic proteins has been suggested to have a role in DNA transcription. Many of the non-histone acidic proteins are phosphoproteins (21). Characterization of the protein kinases which are involved in the phosphorylation of these acidic proteins has recently been reported (22,23).

We have considered the possibility that nuclear protein kinases may play a role in RNA synthesis by a phosphorylation of nuclear RNA polymerases. We wish to report the effects of nuclear protein kinases on polymerases I & II.

METHODS

Rat liver nuclear protein kinases were obtained by the method of Ruddon et al. and Takeda et al. (22,23). Four distinct protein kinases were isolated by chromatography by phosphocellulase chromatography. The protein kinases were not cyclic AMP dependent, phosphorylated casein but did not utilize histone as substrate. Kinase I was used in these studies. Assay conditions were: 50 mM Tris HCl pH 7.5, 0.2 M NaCl, 10 mM $MgCl_2$, 200/μg casein and 0.1 mM γ-^{32}P-ATP.

DNA-dependent RNA polymerase I (nucleolar) and II (nucleoplasmic) were isolated by the method of Roeder & Rutter (24). The assay mixture contained: 0.1 M Tris HCl pH 7.8, 0.8 mM ATP-GTP-CTP, 1 μm H-UTP, 4 mM cercaptoethanol, 2mM $MgCl_2$, 4 mM Mn Cl_2, 20 μg DNA and enzyme.

Polyacrylamide gel electrophoresis was carried out as described by Ornstein (25) and Davis (26).

γ-^{32}P-ATP and γ^{32}-P-GTP were prepared by the method of Glynn & Chappell (27).

RESULTS

The stimulation of RNA synthesis by kinase I is shown in Table 1. The stimulation is optimal with the use of native calf thymus or rat liver DNA. Kinases II, III & IV did not stimulate RNA synthesis. The kinases did not contain RNA polymerase activity. The stimulation by kinase I does not appear to require C-AMP.

TABLE 1

Stimulation of RNA synthesis by protein kinase I

Addition	UMP Incorporation (pmoles/20 min)
None	.056
Polymerase I	0.8
Polymerase I + kinase	5.2
Polymerase I + kinase + C-AMP	5.0
Polymerase II	2.0
Polymerase II + kinase	6.4
Polymerase II + kinase + C-AMP	5.6
Kinase	0.07

The reaction mixture is essentially the same as described in methods except as follows: Polymerase I activity was assayed with $MgCl_2$ and no added salt. Polymerase II assayed with both $MgCl_2$, and $MnCl_2$ with 0.2 M $(NH_4)_2SO_4$. 2 μM cyclic AMP, 8 μg kinase I. Temp 37° for 20 min.

A similar degree of stimulation of RNA synthesis by polymerase I & II was observed with a protein kinase isolated from bovine heart as described by Erlichman et al. (28).

The stimulation of RNA synthesis by kinase I suggests that phosphorylation of RNA polymerase may be occurring. This possibility was tested employing $\gamma-^{32}P$-ATP as phosphate donor. The formation of acid-precipitible ^{32}P was determined following the incubation of polymerase I & II with kinase I. The results are shown in Table 2.

TABLE 2

Phosphorylation of polymerase I & II by kinase I

Additions	^{32}P Incorporated (cpm)
None	150
Protein kinase	350
Poly I	3280
Poly I + kinase	6175
Poly I + kinase + C-AMP	9000
Poly II	0
Poly II + kinase	3000
Poly II + kinase + C-AMP	6530
Protein kinase + casein	66000
Kinase + histone	1200

The reaction mixture was identical to that described in methods except as follows: 10 µg kinase I, 2 µM cyclic AMP, 200 µg casein, 400 µg histone (lysine rich).

Substantial amounts of ^{32}P appear to be incorporated into TCA precipitable material. Polymerase I binds substantial amounts of ^{32}P without added kinase. This was found to be related to the fact that polymerase I preparations contain kinase activity as evidenced by the fact that casein added to polmerase I is greatly phosphorylated. (not shown). The polymerase I & II phosphorylation by kinase I is partly C-AMP dependent. Negligible phosphorylation of histone occurs.

RNA polymerase phosphorylation was further investigated
by incubating intact rat liver nuclei in the presence of
0.1 mM dibutryl C-AMP and γ-^{32}P-ATP followed by isolation
of polymerase I & II as described in methods. The resolution of these phosphorylated polymerase is shown in Figure 1.

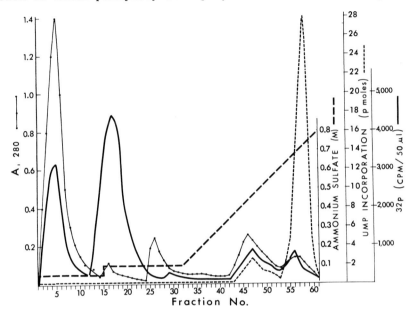

Figure 1.   Resolution of phosphorylated polymerase
I & II.   Assay conditions for RNA synthesis were
identical to those described in Methods.

Figure 1 shows that polymerase I activity corresponds
to ^{32}P.   Other protein peaks which do not contain polymerase activity are also phosphorylated.   The ^{32}P counts do
not correspond exactly to the polymerase II activity.   This
may be due to the fact that Polymerase II exists in two
forms.   Recently, Weaver, et al. have shown that two form
II polymerase enzymes are present in rat liver preparation,
one with a molecular weight (M.W.) structure (190,000),
(150,000), (35,000), (28,000) and another with a structure
(170,000), (150,000), (35,000) and (25,000) (29).   It is
possible that only one form of polymerase II is phosphorylated.

Further studies of ^{32}P-labeled polymerase was made by polyacrylamide gel electrophoresis of the radioactive polymerase I. Unstained gels were sliced and soaked in TGMED for polymerase and ^{32}P elution. The results of this experiment are shown in Table 3.

TABLE 3

Polyacrylamide gel of ^{32}P-polymerase

Slice #	UMP Incorp (CPM)	^{32}P (cpm)
1	158	400
2	150	462
3	180	790
4	160	478
5	175	431
6	180	469
7	160	472
8	200	490
9	350	490
10	200	486
11	160	370
12	550	1680
13	176	439
14	177	380
15	186	456
16	201	382
17	223	287
18	265	400
19	280	301
20	159	236

The reaction mixture was identical to that described in methods except that Mn Cl$_2$ was omitted.

Slice # 12 contained both the RNA polymerase activity and ^{32}P. This experiment was necessary in view of the fact that RNA polymerase I is only partially purified and is contaminated with other proteins after staining of gels with Coomassie Brilliant Blue.

DISCUSSION

The results in this communication show that RNA polymerase I & II from rat liver nuclei are stimulated by a
nuclear protein kinase I. Preliminary studies using column
chromatography and gel electrophoresis suggest that a phosphorylation of these partially purified enzymes may be
occurring.

In general protein kinases have a broad substrate
specificity and phosphorylation in vitro of an acceptor
such as RNA polymerase does not necessarily imply that
this occurs in vivo. However, the fact that kinase I
appears to phosphorylate RNA polymerase I & II in experiments using intact nuclei suggests that a phosphorylation
nay indeed be occurring in vivo as well.

Gene activity in mammalian systems may be controlled
by histone phosphorylation (derepression) as well as by
polymerase phosphorylation.

Acknowledgement
I wish to acknowledge the skillful technical assistance
of Miss Candance Winning.

REFERENCES

1. Cashel, M., and Gallant, J. Nature 221(1969)839

2. Gilbert, W. and Muller-Hill, B. Proc Nat Acad
Sci. U.S.A. 58(1967)2415

3. Zubay, G., Schwartz, D. and Beckwith, J. Proc
Nat Acad Sci. U.S.A. 66(1970)104

4. Emmer, M., deCrombrugghe, B., Pastan, I. and
Perlman, R. Proc Nat Acad Sci U.S.A. 66(1970)480

5. Burgess, R.R., Travers, A.A., Dunn, J.J. and
Bautz, E.K.F. Nature 221(1969)43

6. Davison, M., Pilarski, L.M. and Echols, H. Proc
Nat Acad Sci U.S.A. 63(1969)168

7. Roberts, J.W. Nature 224(1969)1168

8. Chelala, C.A., Hirshbein, L. and Torres, H.N. Proc Nat Acad Sci U.S.A. 68(1971)152

9. Walter, G., Seifert, W. and Zillig, W. Biochem Biophys Res Commun 30(1968)240

10. Martelo, O.J., Woo, S.L.C., Reimann, E. M. and Davie, E.W. Biochemistry 9(1970)4807

11. Walsh, D.A., Perkins, J.P. and Krebs, E.G. J Biol Chem 243(1968)3763

12. Reimann, E.M., Walsh, D.A. and Krebs, E.G. J. Biol Chem 246(1971)1936

13. Chamberlin, M. and Berg, P. Proc Nat Acad Sci U.S.A. 48(1962)81

14. Richardson, J. P. Proc Nat Acad Sci U.S.A. 55(1966) 1616

15. Weber, K. and Osborn, M. J Biol Chem 244(1969)4406

16. Fisher, E.H., Graves, D.J., Snyder-Crittenden, E.R. and Krebs, E.G. J Biol Chem 234(1949)1698

17. Reiman, E.M., Brostrom, C.O., Corbin, J.D., King, C.A. and Krebs, E.G. Biochem Biophys Res Commun 42(1971)187

18. Tao, M.M., Salas, M.L. and Lipmann, F. Proc Nat Acad Sci U.S.A. 67(1970)408

19. Langan, T.A. Proc Nat Acad Sci U.S.A. 64(1969)1276

20. Pogo, B.G.T., Pogo, A.O., Allfrey, V.G. and Mirsky, A.E. Proc Nat Acad Sci U.S.A. 59(1968)1337

21. Kleinsmith, L.J., Allfrey, V.G. and Mirsky, A.E. Proc Nat Acad Sci U.S.A. 55(1966)1182

22. Takeda, M., Yamamura, H. and Ohga, Y. Biochem Biophys Res Commun 42(1971)103

23. Ruddon, R.W. and Anderson, S.L.   Biochem Biophys Res Commun 46(1972)1499

24. Roeder, R.G. and Rutter, W.J. Nature 224(1969)234

25. Ornstein, L. Ann N.Y. Acad Sci 121(1964)321

26. Davis, B.J. Ann N.Y. Acad Sci 121(1964)404

27. Glynn, I.M. and Chapell, J.B. Biochem J 90(1964) 147

28. Erlichman, J., Hirsch, A.H. and Rosen, O.M. Proc Nat Acad Sci U.S.A. 68(1971)731

29. Weaver, R.F., Blatti, S.P. and Rutter, W.J. Proc Nat Acad Sci U.S.A. 68(1971)2994

DISCUSSION

J.P. MILLER: You report here a non-cyclic AMP dependent protein kinase as have many other people. The question I have, and I would like to address it to any speaker here, is what do you think about the possibility that all protein kinases are cyclic AMP dependent and that you just isolated it in a cyclic AMP independent form?

O.J. MARTELO: I do not know the answer to that. It is possible but we have not done any studies to prove this one way or the other. Perhaps people that have been doing more work on the regulatory subunits and cyclic AMP binding of the protein kinases may have more intelligent comments than I may be able to give.

J.P. MILLER: I was thinking particularly in terms of Dr. Rosen's work where after dissociation the regulatory subunit quite possibly stays on the membrane while the catalytic subunit is solubilized and goes into the cytoplasm. Even though the isolated enzyme is cyclic-AMP independent it could be that it is cyclic-AMP dependent in vivo.

T.A. LANGAN: I think there is evidence that there are

protein phosphokinases that are not cyclic AMP dependent
and probably not related at all to the cyclic AMP control
systems. And of course, the first one you might think of
is phosphorylase kinase but in addition we have investigated an enzyme that puts phosphate in a different location
in lysine-rich histone and in colaboration with Dr. Walsh
and Dr. Krebs, we have investigated the interaction of that
protein kinase with regulatory subunit from skeletal muscle
and liver cyclic AMP dependent protein kinases and there
is no interaction. So, this question of whether an isolated cyclic AMP independent protein kinase is related to the
cyclic AMP dependent type is a question you can answer
rather directly experimentally by studying the interaction
of that enzyme with the regulatory subunit. I feel that
a little more emphasis has to be given in people's minds
to the possibility that protein phosphorylation is a regulatory mechanism, that probably goes beyond cyclic AMP
systems and has many more functions than we are even aware
of so far.

E.G. KREBS: I would strongly second Dr. Langan's remarks.

T. SODERLING: In your abstract you allude to the possible
presence of a protein kinase in E. coli by stating that
partially purified RNA polymerase apparently has protein
kinase activity. I wonder if you would expand a little on
this?

O.J. MARTELO: When we first began trying to see the effect
of the rabbit muscle protein kinase on RNA polymerase (RNA
synthesis) and on phosphorylation of the enzyme, we found
that early in the purification of the enzyme there is no
stimulation of RNA synthesis by the muscle protein kinase.
When we tried to do phosphorylation experiments the experiments were meaningless because of the fact that the polymerase by itself bound significant amounts of ^{32}P from
gamma labeled ATP^{32} in the presence of cyclic AMP. This
is the only evidence that I have from our laboratory that
there appears to be a cyclic AMP dependent protein kinase.
Now this binding of ^{32}P by polymerase completely disappears
once the enzyme is pure. So I think that there is a protein kinase in E. coli, but for some reason we have not
been able to purify it.

V.G. ALLFREY: One of the questions I have deals with your results showing that cyclic AMP had no effect upon the protein kinase activity, yet when it was added to the RNA synthesizing system, it did stimulate RNA synthesis. The question is whether you may be dealing with a protein such as the cyclic AMP-receptor protein described by Dr. Pastan and by Dr. Zubay, which can influence the rate of RNA synthesis by conformational changes which accompany cAMP-binding rather than by modifying protein kinase activities.

O.J. MARTELO: You are referring to E. coli experiments? We have found that we can phosphorylate the enzyme and it seems to require cyclic AMP but the bulk of the stimulation of RNA polymerase by the muscle protein kinase does not really require cyclic AMP. We do not know why this is. We have thought that perhaps the DNA or even the enzymes might contain something that activates the protein kinase. We are not sure what is happening because our reaction mixture for phosphorylation is different as from our reaction mixture for RNA synthesis. Our reaction mixture for RNA synthesis contains DNA and whether there is some contaminant there or not we are not sure. Now, as far as the cyclic AMP receptor protein as described by Zubry and Emmer is concerned, it has been reported that the cyclic AMP receptor protein does not contain kinase activity. It appears to bind to the DNA template. I do not know if there is any correlation between the cyclic AMP receptor protein and cyclic AMP protein kinase from E. coli. I think that cross experiments between the E. coli and the mammalian kinases are not quite valid.

E.M. JOHNSON: You said that your kinase I was not cyclic AMP dependent, but in one of your slides it appears that there was a considerable effect on phosphorylation by cyclic AMP. Could you explain that please?

O.J. MARTELO: I can not explain why using casein it does not require cyclic AMP, but when we use polymerase as a substrate, then we see a further increase in incorporation.

E.G. KREBS: I would like to comment briefly on both the calcium dependent protein kinase and also the cyclic AMP dependent protein kinases. We appear to have a phenomenon in which the response of the enzyme to its effectors seems

215

to vary a great deal with the nature of the substrate.
With the cyclic AMP dependent protein kinase a part of this
variation may be explained by the observations in Greengard's laboratory and also in our laboratory that basic
proteins dissociates the cyclic AMP dependent protein
kinase. Whether or not that is the entire explanation, I
am not sure.

I would like to make one last comment. If one looks
at the total amount of acylphosphate which probably represents intermediates in enzymatic reactions found in E. coli
in comparison to the amount found in serine phosphate, this
ratio of acylphosphate to serine phosphate in E. coli is
much higher than in mammalian cell. In other words,
mammalian cells are loaded with serine phosphate. In
E. coli certainly most of the protein bound phosphate represents intermediary phosphates. So that on this basis
there would be some justification for at least condluding
that protein phosphorylation in E. coli is relatively unimportant compared to protein phosphorylation in mammalian
cells. I do not mean to say that I doubt the significance
of the phosphorylation that you have shown here.

PHOSPHORYLATION OF NUCLEAR PROTEINS
AT TIMES OF GENE ACTIVATION

V.G. ALLFREY, E.M. JOHNSON, J. KARN and G. VIDALI
The Rockefeller University
New York

Abstract: Cell nuclei from different animal tissues contain sets of acidic proteins which appear to play a role in the regulation of transcription. Many of these proteins are phosphorylated and de-phosphorylated in the cell nucleus.

The nuclear acidic proteins show characteristic distributions in the various cell types of the adult organism and they vary during embryonic development of particular cell lines. Erythroid cells from different developmental stages of the duck differ in the nature of major gene products such as hemoglobin and histone F2C and they also show specific differences in their nuclear acidic protein complements.

The uptake of radioactive phosphate into nuclear acidic proteins correlates closely with gene activation in a number of cell systems. This is evident in liver cells stimulated by cortisol and by cyclic AMP. Cortisol administration leads to a complex series of time-dependent changes in the pattern of nuclear protein phosphorylation, with major stimulations evident within 15 minutes. Different nuclear proteins respond differently at different times after injection of the hormone. The effects of dibutyryl cyclic AMP differ from those of cortisol but are also evident within 15 minutes.

The phosphorylation of the nuclear acidic proteins is modified by histones. Different histone fractions affect the phosphorylation of individual nuclear acidic proteins in a characteristic manner. One of the nuclear phosphoproteins acts as a histone deacetylase, suggesting interlocking control mechanisms in nuclear protein metabolism.

Studies of nuclear acidic proteins during the cell cycle in synchronized HeLa cells indicate that phosphorylation is maximal in the S̲ phase and in the period encompassing late M and early G̲₁. Phosphorylation of the nuclear proteins is minimal in the late S̲ and G₂ phases when RNA synthesis is also suppressed.

The binding of some nuclear phosphoproteins to DNA has been studied by DNA-cellulose column chromatography. There is evidence for both species and sequence specificity.

The nuclear acidic protein fractions include components which stimulate transcription from free DNA in vitro if the DNA and the proteins are derived from the same species. Enzymatic removal of the phosphate groups abolishes the capacity of the phosphoproteins to promote RNA synthesis.

It is proposed that phosphorylation of the acidic nuclear proteins constitutes part of the mechanism for positive control of transcription. Since the phosphorylation of many of the nuclear acidic proteins is stimulated by cyclic AMP, the effect of cAMP on transcription may well be direct.

I. INTRODUCTION

In both eukaryotic and prokaryotic cells the utilization of DNA as a template for RNA synthesis requires both a restriction and a selective activation of particular genetic loci. The mechanisms through which such control is achieved involve, at least in part, the participation of DNA-associated proteins which influence the structure of the genetic material and its interactions with RNA polymerases.

Examples of DNA-associated regulatory proteins in bacteria include the repressor protein which inhibits transcription of the lac operon (1, 2) and the cyclic AMP-receptor protein which promotes transcription of gal messenger-RNA (3, 4). The first is a relatively acidic protein with a high specificity for particular nucleotide sequences in DNA, while the latter is a highly basic protein with less discrimination in its DNA-binding properties.

In higher organisms the corresponding elements of genetic control - i.e., suppression of template activity of most of the DNA, and activation of RNA synthesis at particular genetic loci - also require the intervention of proteins with different properties, DNA-binding affinities, and contrasting effects on transcription. The proteins concerned include the histones - the basic, suppressive, structural components of chromatin - as well as more acidic proteins which show strong indications of involvement in the positive control of RNA synthesis at specific sites on the genome.

The present report concentrates on the latter class of nuclear acidic proteins, emphasizing their diversity in different cell types, as well as their DNA-binding properties, stimulatory effects on transcription, and altered metabolism in hormone activated tissues. Particular attention will be paid to the post-synthetic modification of such proteins by enzymatic phosphorylation and de-phosphorylation of seryl and threonyl residues in their polypeptide chains. These changes, which ultimately affect the nature and strength of the interactions between the proteins and DNA, offer informative clues to the molecular events underlying the "scanning" of the genome andthe selective activation and repression of particular cistrons in different cells at different times.

"Acidic"Proteins of the Nucleus -   In recent years there has been a revival of interest in the non-histone proteins of the cell nucleus, particularly in those aspects of their metabolism related to genetic control. (The chemistry and function of the nuclear proteins have been the subject of several recent reviews (5 - 8).)

The view that acidic nuclear proteins play a role in the regulation of transcription stems from a wide variety of observations which can be summarized briefly -

1 - The non-histone proteins are present in higher concentrations in the chromatin of metabolically-active cells than in chromatin from inert cell types (9, 10),

2 - They are preferentially localized in those regions of the chromatin that are most active in RNA synthesis (11-16),

3 - As would be expected for proteins involved in differential gene expression, they are distributed characteristi-

cally in the nuclei of different somatic tissues (17 - 21),

4 - The nature and amount of the non-histone chromosomal proteins changes during differentiation and aging of a particular cell type (22 - 25),

5 - The synthesis (26 - 2 8) and phosphorylation (29 - 3 2) of different non-histone nuclear proteins are modified at times of gene activation by estrogens, androgens and other steroid hormones and peptide hormones,

6 - Their phosphorylation is selectively stimulated by cyclic AMP (33),

7 - Their metabolism is altered during gene activation induced by drugs such as phenobarbital (34) and isoproterenol (35),

8 - Increased synthesis of the nuclear acidic proteins is observed during the release of WI-38 cells from contact inhibition (36),

9 - Increased phosphorylation of the nuclear phosphoproteins is evident within minutes after lymphocytes are stimulated by mitogenic agents such as phytohemagglutinin (37),

10 - Certain nuclear phosphoproteins combine selectively with the DNA of the species of origin (17, 18, 38),

11 - They stimulate the rate of transcription from the appropriate DNA templates in vitro (39 - 4 2),

12 - The nature of the RNA synthesized by isolated chromatin fractions (as judged by RNA/DNA hybridization) depends on the nature of the non-histone proteins present during the RNA polymerase assay (43 - 49).

Nuclear Protein Phosphorylation -   Early tracer studies of nuclear metabolism had indicated that some nuclear proteins are rapidly phosphorylated in vivo (50, 51). The isolation of nuclear phosphoprotein fractions and studies of the enzymatic basis of their phosphorylation were initiated by T.A. Langan in 1964 (5 2). Subsequent work by Kleinsmith and Allfrey established that both histones and non-histone proteins contain phosphoseryl residues, and that most of the phosphate incorporated into nuclear proteins by non-dividing cell nuclei is recovered as phosphoserine in the acidic protein fraction (53 - 55). It was shown that the phosphorylation of the acidic nuclear proteins is reversible, in the

sense that phosphate groups once incorporated into seryl
and threonyl residues can be enzymatically removed without
degradation of the polypeptide chain (5 3-55). Thus, both
phosphorylation and de-phosphorylation take place on intact
polypeptide chains in the cell nucleus.

The nuclear phosphoproteins comprise a complex mixture
differing in electrophoretic mobility and molecular weight
(56, 17, 18). On the average, the phosphorus content is
about 1.0 - 1.3 % by weight.

A relationship between phosphorylation and genetic activity is suggested by a number of observations -

1 - Phosphate uptake into lymphocyte nuclear proteins is
increased prior to the increase in RNA synthesis during gene
activation by phytohemagglutinin (37),

2 - The phosphoprotein content of active (euchromatic)
chromatin fractions greatly exceeds that of the inactive
(heterochromatic) fractions (11),

3 - The phosphoprotein complement of avian erythrocyte
nuclei decreases during the course of erythrocyte maturation
and the concomitant repression of RNA synthesis (2 3),

4 - The ability of nuclear acidic proteins to stimulate
transcription in vitro increases with increasing phosphorylation (60) and decreases upon enzymatic removal of their
phosphate groups (32, 42).

It has been proposed that the rapid phosphorylation of
the nuclear acidic proteins represents an important aspect
of positive genetic control (61, 62). The ready reversibility
of the process fits a cyclical mechanism for the binding and
release of regulatory proteins which direct the attachment of
RNA polymerases to specific sites for RNA chain initiation
on the chromatin (31). The view that genes are turned-on
and turned-off by interaction with control proteins whose
conformation and DNA-binding properties are alterable by
phosphorylation has received further confirmation in the
experiments now to be described.

II. Changes in Nuclear Phosphoproteins in Development -
A method for the preparation of the nuclear phosphoproteins based on their localization in chromatin, and their
characteristic solubility properties has recently been

described (26, 17, 18). Nuclei are first purified by centrifugation through sucrose density barriers (18). A differential extraction of the nuclear proteins is then carried out, using 0.14 M NaCl to remove saline-soluble components, 0.25 N HCl to remove histones, and phenol to solubilize many of the acidic proteins of the residue. The proteins in the phenol phase are dialyzed against a series of urea-containing buffers which restore them to the aqueous phase for further characterization. The separation and characterization of individual proteins in the extract is achieved by polyacrylamide gel electrophoresis in the presence of sodium dodecylsulfate (18). A complex banding pattern is obtained in which the distance of migration of individual proteins can be correlated with their molecular weights (Figure 1). The relative amounts of individual components have been determined in two ways : by staining of the protein bands with Amidoblack 10B, followed by quantitative densitometry and computer analysis of the densitometer tracing, or by cutting out individual bands and extracting the dye from each slice in dimethylsulfoxide. In labelling experiments the ^3H and ^{14}C content of the protein in each gel slice was measured by scintillation spectrometry, usually after combusting the slice in a sample oxidizer and collecting the radioactive CO_2 and water in separate vials. The ^{32}P content of the nuclear phosphoprotein bands was determined by quantitative autoradiography, or by direct counting of each slice after solution in H_2O_2 (33). (The phenol isolation procedure offers particular advantages in the study of nuclear protein phosphorylation because it minimizes contamination by radioactive nucleic acids which remain in the aqueous phase during the partition step. Chemical and isotopic tests for RNA and DNA contamination of the final phosphoprotein preparations have proven negative (18).) It is a surprising fact that the phenol-treated proteins are not irreversibly denatured during their isolation; certain enzyme activities, including protein kinase activity, are restored by gradient dialysis against urea-salt solutions in which the salt concentration is progressively reduced and the urea is finally eliminated (18). This treatment also restored specific

DNA-binding properties to the protein fraction (17, 18).

It has been shown that the phenol-soluble nuclear protein fraction is heterogeneous and that the electrophoretic banding patterns differ from one cell type to another (18). Antigenic differences in nuclear acidic protein from diverse tissues have also been described (2 1).

If such proteins play a role in differential transcription of the genome it might be expected that their proportions and metabolism would be altered in the course of cell differentiation. This question has been examined in avian erythroid cells during embryonic development. On the premise that the "primitive" red cell series of the very early duck embryo would differ in transcriptional control mechanisms from erythroid cells of the later embryo and adult stages, we have compared the acidic proteins of nuclei isolated from the different cell types. It is known that these cell types differ in the nature of their hemoglobins (63) and in their contents of the erythrocyte-specific histone F2C (25). The red cells of the 7-day embryo, for example, carry the fetal hemoglobins, Hb III and Hb IV, while erythrocytes of the later, definitive red cell series carry the adult hemoglobins, Hb I and Hb II. The cells also differ in their total Hb content, in their capacity for RNA and DNA synthesis, and in details of morphology, particularly with regard to heterochromatinization of the nucleus.

The acidic nuclear proteins were prepared from nuclei isolated from erythroid cells of the 7-day embryo, the 14-day embryo, and the mature adult white Peking duck. The electrophoretic banding pattern of the proteins of the 7-day embryonic red cells are shown in Figure 1A; the corresponding patterns for adult red cells are shown in Figure 1B. The corresponding densitometric tracings are plotted to scale above the photographs of the gel patterns. Estimated molecular weights are indicated above the tracings.

A number of differences exist between the nuclear proteins of the embryonic and adult red cells. Major bands of molecular weights 130, 000 and 136, 500, evident in the 7-day embryo, are greatly reduced in the adult, as is the band at molecular weight 110, 000. Major decreases are also

evident in bands at 68,000 and 82,000. The peaks at
46,000 and 99,000 remain prominent in the adult erythrocyte.
Shifts in relative concentration and intensity occur in multiple bands in the molecular weight ranges 27,000-38,000
and 51,000-58,000, with emergence in the adult of prominent bands at 34,000 and 51,000. (Control experiments, in
which cytoplasmic proteins and membrane proteins were extracted and analyzed in the same way have established that
none of these changes represent variable amounts of contamination by non-nuclear proteins.)

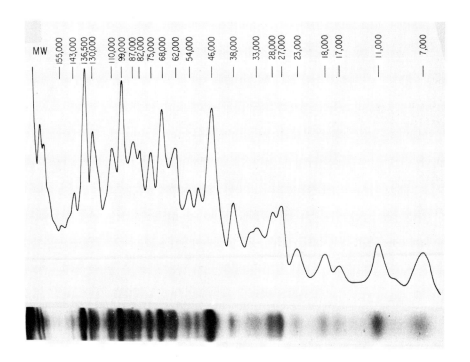

Figure 1A. Acidic nuclear proteins from erythroid cells of the
7-day embryo of the white Peking duck. The molecular weights
of the major protein bands separated by SDS-polyacrylamide
gel electrophoresis are indicated above the gel pattern.

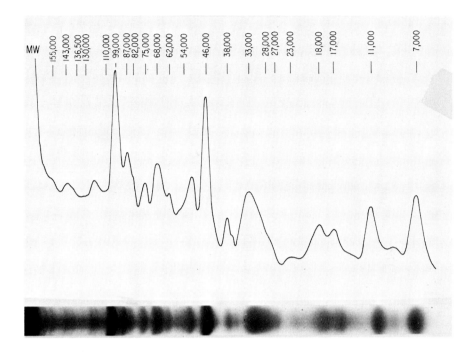

Figure 1B. Acidic nuclear proteins of erythrocytes from mature adult white Peking ducks. The molecular weights of the major protein bands separated by SDS-polyacrylamide gel electrophoresis are indicated above the gel pattern.

Thus, it is clear that erythroid cells from different developmental stages differ in their complement of acidic nuclear proteins. The changes correspond to the disappearance of the "primitive" red cell series and the emergence of the second, definitive red cell series carrying the adult hemoglobin types. In a speculative vein, it may be proposed that some of the changes in nuclear acidic proteins reflect the altered genetic controls which affect the synthesis of different messenger-RNA's.

225

Although the selective loss of certain nuclear acidic proteins (e.g. of molecular weights 130,000 and 136,500) correlates with the disappearance of the fetal hemoglobins, it should be pointed out that these experiments do not establish a direct cause-and-effect relationship between the changes in the nuclear acidic proteins and the nature of the hemoglobins synthesized in embryonic and adult red cells.

It is evident that cells of similar morphology and function obtained at different stages of embryogenesis and maturation can differ both in the nature of their major product (63) and in their complements of acidic proteins believed to play a role in selective transcription. Similar differences in the nature and distribution of acidic nuclear proteins in avian erythroid cells have also been detected at different stages of red cell maturation in the adult (24). The results are consistent with the view that the nuclear acidic proteins are involved in the programming of differential gene activity in the development and maturation of individual cell types.

III. Nuclear Phosphoprotein Metabolism and Gene Activation by Hormones -

The uptake and release of phosphate by nuclear phosphoproteins is a major aspect of their metabolism (50 - 56). Studies of ^{32}P-orthophosphate incorporation into individual nuclear acidic proteins separated by disc gel electrophoresis have indicated clear differences in the extent and kinetics of phosphorylation of different nuclear proteins from the same tissue. This is shown for rat liver phosphoproteins in Figure 2A which plots the ^{32}P-activity of individual protein bands as a function of their mobility in SDS-polyacrylamide gels. The corresponding labelling pattern for kidney nuclear phosphoproteins in the same experimental animals is shown in Figure 2B (18). There are obvious differences in ^{32}P distribution in the analytical gels from the different tissues. Thus the phosphorylation pattern as well as the protein distribution pattern indicate tissue specificity.

The synthesis and the phosphorylation of individual nuclear acidic proteins are subject to hormonal control. In the liver, cortisol selectively stimulates the synthesis of

a non-histone nuclear protein of molecular weight 41,000
(26). Glucagon administration leads to an increased uptake
of radioactive leucine into two other liver nuclear proteins of
molecular weight 60,000 and 80,000, respectively (28).
The synthetic response to these hormonal stimuli is compar-
atively slow, requiring several hours for its full expression
(26,28). Much more rapid responses are evident in [32]P
uptake into the nuclear acidic proteins.

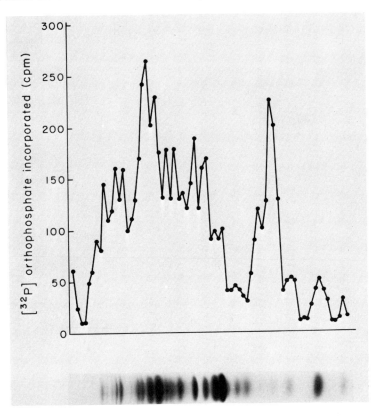

Figure 2A. Distribution of ^{32}P in rat liver nuclear
phosphoproteins after 90 minutes' phosphorylation
in vivo. The proteins were separated by SDS-poly
acrylamide gel electrophoresis.

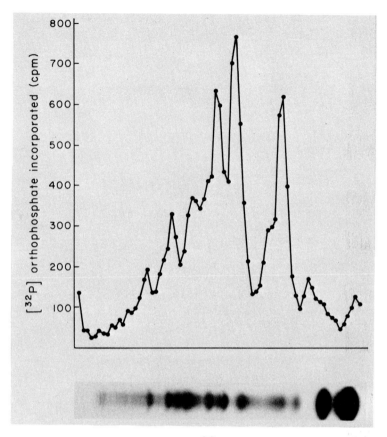

Figure 2B. Distribution of ^{32}P in rat kidney nuclear phosphoproteins after 90 minutes phosphorylation *in vivo*.

The phosphorylation of the nuclear phosphoproteins of the liver is stimulated within 5 minutes after administration of cortisol (31, 32). Studies of the kinetics of phosphorylation of individual protein bands in the liver phenol-soluble fraction show that the time courses of phosphate uptake and release are highly complex. They differ from one protein to another, and the patterns of phosphorylation vary with time after hormone administration. Some indication of the complexity is shown in Figures 3A and 3B, which compare the specific ^{32}P-activities of the phosphoproteins of control and cortisol treated rats at 15 minutes and at 120 minutes after hormone injection. Stimulations of ^{32}P-uptake into proteins of high molecular weight are evident at early times but not later (compare Figures 3A and 3B). Several proteins of lower molecular weight (higher mobility) show a greater stimulation of ^{32}P-uptake at later times after hormone administration. (It should be pointed out that the differences in ^{32}P-uptake are not likely to be due to fluctuations in the nuclear ATP "pools" which would be expected to shift the labelling of all the phosphoproteins in a parallel fashion.)

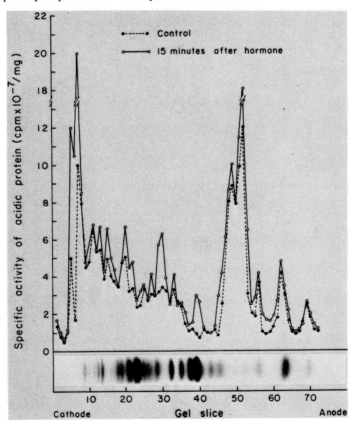

Figure 3A. Comparison of ^{32}P distribution in acidic nuclear proteins of control and cortisol-treated rat livers; 15 minutes after hormone injection.

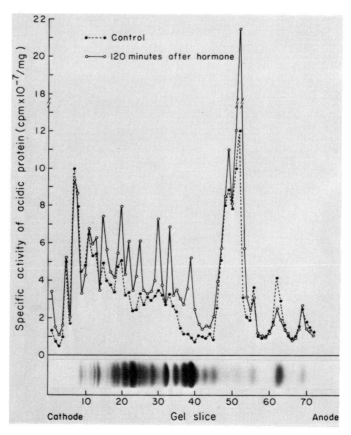

Figure 3B. Comparison of ^{32}P distribution in acidic nuclear proteins of control and cortisol-treated rat livers; 120 minutes after hormone injection.

Whether these early and late responses in phosphorylation are causally related to the known programming of synthesis of different RNA types in the cortisol-treated liver (64, 65) remains to be determined.

The stimulation of protein phosphorylation by hormones is not limited to the acidic proteins of the nucleus. An increased phosphorylation of liver histones occurs after the administration of cortisol (66), glucagon (67), insulin (67) and cyclic AMP (68). It follows that hormone-induced modifications of the nuclear proteins are likely to affect the physical state of the chromatin as well as its function in the target cell.

IV. Cyclic AMP Effects on the Phosphorylation of Acidic Nuclear Proteins in the Liver –

The administration of dibutyryl cyclic AMP to adrenalectomized rats alters the pattern of phosphorylation of the acidic nuclear proteins of the liver (33). It also augments the uptake of phosphate into the liver histones (68, 33).

The effects of cyclic AMP on the phosphorylation of nuclear acidic proteins are rapid and complex. Within 15 minutes, an increased incorporation of ^{32}P-orthophosphate is observed, the average increment amounting to about 30% over control values. The average is misleading, because some of the protein bands show increases in specific activity of over 200% (Figure 4). Proteins in the molecular weight ranges 15,000-40,000 and 60,000-85,000 account for the most significant increases in phosphorylation. Selectivity in the effects of cyclic AMP is evident in that many protein bands are not stimulated. Particularly, a peak at MW 57,000 showing the highest total phosphate incorporation has the same specific activity in control and treated animals. Certain of the higher molecular weight peaks are also unaffected by cyclic AMP treatment.

In confirmation of the earlier results by Langan (68) we observed that the phosphorylation of the liver histones increases in cyclic AMP-treated animals. An eleven-fold stimulation of phosphorylation of histone fraction F1 could be detected 15 minutes after injection of dibutyryl cyclic AMP.

Definite effects of cyclic AMP on the phosphorylation of nuclear acidic proteins can also be demonstrated in isolated rat liver nuclei, the maximum stimulation occurring at about 5×10^{-6}M cyclic AMP. (When the phosphorylated nuclear proteins are extracted and separated by SDS-polyacrylamide gel electrophoresis, the changes due to cyclic AMP are confined to molecular weight regions much higher than those of the histones, thus eliminating the possibility that histone contamination of the acidic protein fraction could be responsible for the increased phosphorylation observed in the

231

presence of cyclic AMP.) While the magnitude of the stimulation observed in isolated nuclei is considerably lower than
that observed when dibutyryl cyclic AMP is administered in
vivo , the effects are selective for specific acidic proteins
in both cases. For example, cyclic AMP stimulates phosphorylation of proteins in the 70,000 MW range in vivo and
in vitro. However, the labelling pattern of other acidic
proteins phosphorylated in isolated nuclei and separated by
gel electrophoresis differs in some respects from the pattern
obtained when the acidic proteins are phosphorylated in vivo.
Some of the differences reflect the probable loss of nuclear
kinases during the isolation of the nuclei in aqueous media,
while others may be due to the short incubation times (4
minutes) and unphysiological conditions employed in the
in vitro experiments.

The ^{32}P labelling experiments demonstrate the rapidity
with which cyclic AMP, an important mediator of the action
of many hormones, stimulates the phosphorylation of certain
nuclear acidic proteins both in vivo and in vitro. The
results are in accord with the view that the induction of
enzymes (69-72) and the increase in nuclear RNA synthesis
(73) caused by cyclic AMP in the rat liver involve the phosphorylation of the chromosomal acidic proteins as an early
stage in transcriptional control.

V. Histone Effects on Acidic Nuclear Protein Phosphorylation

Recent investigations suggest that histones may play a
role in nuclear protein phosphorylation. For example, the
addition of histones to partially purified calf thymus nuclear
phosphoproteins stimulates their phosphorylation (74).
Added histones have also been shown to effect a dissociation of the regulatory and catalytic subunits of a cyclic AMP-
dependent protein kinase from bovine brain (75).

The effects of histones on the phosphorylation of nuclear
acidic protein has been studied by incubating isolated rat
liver nuclei in the presence of (γ -^{32}P) ATP and increasing
amounts of purified exogenous histone fractions. The radioactivity of the phosphoprotein bands separated by SDS-polyacrylamide gel electrophoresis was subsequently determined.

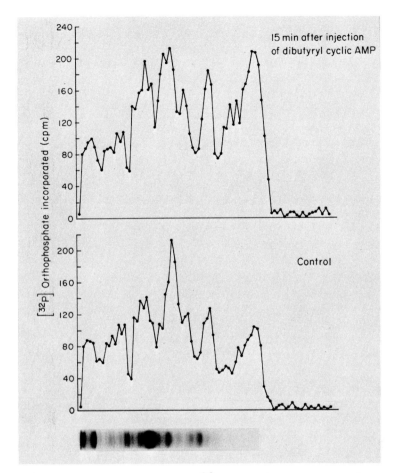

Figure 4. Comparison of 32P distribution in nuclear acidic proteins of control rats and rats receiving dibutyryl cyclic AMP 15 minutes earlier.

(The fact that exogenous histones do enter the isolated hepatocyte nuclei was verified by high resolution autoradiography with the electron microscope, using histones labeled with ^{125}I-iodine (76). It was shown that binding of exogenous histones occurs throughout the nucleoplasm and is not restricted to the nuclear membrane.)

The addition of increasing amounts of histone F2A1 has a differential effect on the phosphorylation of different

233

acidic proteins in the nucleus; inhibition is observed in some cases, stimulation of phosphorylation in others. For example, the phosphorylation of a protein fraction of molecular weight 7,000 is markedly inhibited at low concentrations of F2A1 and virtually abolished at high concentrations (76). Tests to identify the sites of phosphorylation in this acidic protein show that about 70% of the incorporated ^{32}P is recovered as phosphoserine and about 25% as phosphothreonine. Phosphorylation of both amino acids is inhibited by added F2A1.

In contrast, increasing amounts of histone F2A1 stimulate the phosphorylation of a protein fraction of molecular weight 22,000, the stimulation amounting to about 1.7-fold (when 8 mg of histone F2A1 are added to a nuclear suspension containing 15 mg of total nuclear protein). Slight stimulations of ^{32}P-incorporation are also seen in proteins in the molecular weight ranges 40,000-80,000, 100,000 and 170,000.

Different histone fractions differ in their effects on nuclear acidic protein phosphorylation. Comparisons of the very lysine-rich fraction F1 with the arginine-rich F2A1 histone show that both fractions inhibit the phosphorylation of a low molecular weight acidic protein, but each histone stimulates phosphorylation of a different set of nuclear proteins. Histone F2A1 augments phosphate uptake into a protein fraction of approximate molecular weight 22,000, while the most striking effect of histone F1 is a stimulation of phosphorylation of proteins in the molecular weight range 30,000-60,000.

Histones F2A2, F2B, and F3 were also tested for their effects on the phosphorylation of different non-histone nuclear proteins. Histone F2A2 stimulates phosphorylation of proteins in the 30,000-60,000 size range but has little or no effect on phosphate uptake into the low molecular weight fractions. Histones F2B and F3 have only slight effects on ^{32}P incorporation into the non-histone proteins of isolated rat liver nuclei.

In interpreting the results of these experiments, it should be noted that the addition of extra histone to calf thymus chromatin has only minor effects on free DNA-phosphate content and template activity (77). This is consistent with the view that binding of exogenous histones by chromatin fractions involves regions of the DNA which already have some histone (and non-histone) proteins bound to them. There are many indications that histones interact with nuclear acidic proteins (52, 78-82). The formation of complexes with nuclear phosphoproteins is clearly demonstrable in vitro (52, 83) and is sometimes highly selective (83). Such interactions are likely to occur when individual histone fractions are added to isolated hepatocyte nuclei.

The interaction between histones and acidic chromosomal proteins is evident in two ways. The effects on phosphorylation show that different histones may modify the structure and metabolic activity of individual nuclear phosphoproteins, Conversely, it is known that a nuclear phosphoprotein that binds selectively to histones F2A1 and F3 functions as a histone deacetylase (83). This mutual interdependence of enzyme systems regulating the acetylation of histones and the phosphorylation of nuclear acidic proteins may represent an important aspect of interlocking controls in the regulation of genetic activity.

V I. Nuclear Protein Phosphorylation during the Cell Cycle -
    Studies of nuclear protein synthesis in synchronized HeLa cells have established that the amount of protein synthesized, transported, and retained in the acidic protein fraction is greater immediately after mitosis and later in G_1 than in the S or G_2 phases of the cell cycle (84).
    Individual polypeptides separated by SDS-polyacrylamide gel electrophoresis have widely-different rates of synthesis and turnover which depend on the stage in the cycle (84). An initiation of synthesis of the acidic nuclear proteins before the S phase of the cell cycle has been noted in other systems, such as mouse salivary glands stimulated by isoproterenol (85), and in rat uterus stimulated by estrogen (81) or by progesterone (86). Similar observations have been made on fibroblasts released from contact inhibition (36) and in explanted cells of the rat mammary gland (87).
    While stage-specific differences in the synthesis of

235

different proteins have been demonstrated in synchronized HeLa cells (84, 89), little is known about the phosphorylation of the acidic nuclear proteins in relation to the cell cycle. This has been investigated in HeLa cell cultures following synchronization after a double thymidine block (88). The cells were pulse-labelled with ^{32}P-phosphate for 15 minutes at 3 hour intervals during the cell cycle. The chromatin was isolated (89, 90) and the phenol-soluble nuclear protein fraction was prepared in the usual way (18). The banding patterns of the HeLa nuclear acidic proteins at different times in the cell cycle are shown in Figure 5. Superficially, they seem very similar, but a number of differences in relative concentrations of individual bands can be observed. Such differences have been noted before in the non-histone chromosomal proteins of synchronized HeLa cells (91).

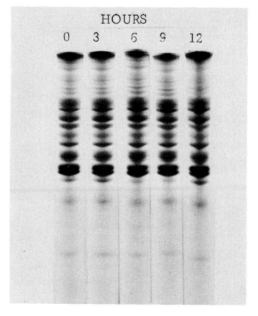

Figure 5. Electrophoretic banding patterns of acidic chromosomal proteins of synchronized HeLa cells at different times in the cell cycle.

The relationship between ^{32}P incorporation into the nuclear phosphoproteins and other parameters of growth is shown in Figure 6. The peak of DNA synthesis occurs at 3 hours and the mitotic index reaches a maximum at 8 hours. There are two peaks in nuclear protein phosphorylation, the first occurring in the S phase and the second in the late M

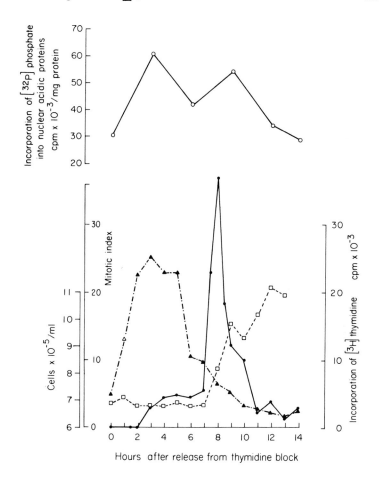

Figure 6. Changes in nuclear protein phosphorylation during the cell cycle in synchronized HeLa cells (upper curve). Other parameters of cell growth plotted include TdR uptake, mitotic index and cell number.

to early G_1 phases of the cycle. Phosphorylation of the nuclear acidic proteins is minimal in the late S and G_2 periods when RNA synthesis is also suppressed.

There are great differences in the extent of phosphorylation of individual acidic nuclear proteins of synchronized HeLa cells. Figure 7 plots the ^{32}P-activity as a function of the electrophoretic mobility (MW) of the proteins isolated from the chromatin during the early G_1 phase. The metabolic heterogeneity of the nuclear phosphoproteins is evident. We are currently investigating how the phosphorylation of individual phosphoproteins varies during the cell cycle and how it relates to changing levels of cyclic AMP in these cells (92). It is significant that the minimum of protein phosphorylation is observed in G_2, when cAMP levels are known to decline (92).

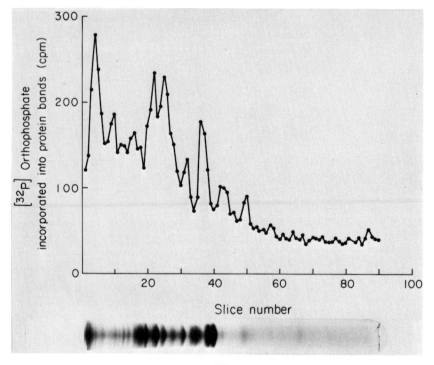

Figure 7. Distribution of ^{32}P in the HeLa cell nuclear phosphoproteins in the early G_1 phase of the cell cycle.

VII. Conclusions - Specificity of Phosphoprotein-DNA
Interactions and Effects on Transcription in vitro -

The interactions between phosphoproteins and DNA
have been investigated in binding experiments using ^{32}P-
labelled phosphoproteins from rat liver and kidney nuclei
and DNA from a number of animal and bacterial species.
Samples of DNA and protein, each dissolved in 2 M NaCl-
5 M urea-0.01 M tris at pH 8.0, were mixed and dialyzed
together against a progression of salt solutions of decreas-
ing concentration. After removal of the urea, the samples
were centrifuged through sucrose density gradients to separate the DNA-protein complexes (17, 18). The specificity
of DNA binding was indicated by the failure of the rat
nuclear acidic proteins to combine effectively with DNA's
prepared from calf thymus, salmon sperm, human placenta,
dog liver, and bacterial sources, while both liver and kidney nuclear proteins did form complexes with rat DNA. A
similar specificity in DNA-binding has been reported by
Kleinsmith, Heidema and Carroll (38) who used DNA-cellulose affinity chromatography to demonstrate that a fraction
of the non-histone protein from rat liver chromatin binds to
rat DNA but not to DNA's from salmon sperm or E.coli.

We have recently investigated the binding of calf thymus
nuclear acidic proteins to DNA-cellulose columns bearing
unique and reiterated sequences of calf thymus DNA which
were separated by hydroxyapatite chromatography (93).
The proteins were eluted in a stepwise salt gradient and
the components in each fraction were analyzed by SDS-polyacrylamide gel electrophoresis. The method separates
proteins of different DNA affinities and size distributions.
It promises to simplify the separation of proteins with
selective affinities for reiterated and unique DNA sequences
(93).

The view that acidic nuclear proteins selectively bind
to DNA in vivo is supported by autoradiographic studies
which show a localization of ^{32}P-labelled proteins along
the salivary gland chromosomes of Sciara larvae (30). The
puffing of particular loci of Drosophila hydei chromosomes
induced by ecdysone or by temperature shifts also correla-

239

tes with increased accumulation of specific acidic proteins in the cell nuclei (13).

A number of studies in this and other laboratories indicate that the nuclear acidic proteins stimulate transcription from free DNA in vitro using either the E.coli RNA polymerase or RNA polymerase II (B) from rat liver (18, 41, 42). The stimulatory effect is species specific because rat proteins fail to promote RNA synthesis in systems templated by DNA's from calf thymus (18, 41), salmon sperm (42), E. coli (41, 42) or Clostridium perfingens (42).

Enzymatic removal of the phosphate groups from the rat liver nuclear phosphoproteins (using E.coli alkaline phosphatase) reduces and eventually abolishes their capacity to promote RNA synthesis under these test conditions (32, 42).

*   *   *

The results support the view that the nuclear phosphoprotein fraction includes components which are involved in the positive control of RNA synthesis. The mechanism by which transcription is selectively facilitated is not known, but it is likely to involve binding of the phosphoprotein to specific nucleotide sequences in the DNA. We propose that phosphorylation is a critical variable in the interactions between the phosphoproteins, DNA, and the RNA polymerases. The reversibility of the phosphorylation of the nuclear proteins in vivo strongly suggests that there is a cyclical binding and release of the regulatory proteins from sites of initiation on the chromosome. Since the phosphorylation of many of the nuclear acidic proteins is stimulated by cyclic AMP, the effect of cAMP on transcription may well be direct.

REFERENCES

1. W.Gilbert and B.Muller-Hill, Proc.Nat.Acad.Sci.U.S. 58 (1967) 2415.
2. A. D.Riggs, R.F.Newby, J.Bourgeois and M.Cohn, J.Mol. Biol. 34 (1968) 365.
3. W.B.Anderson, A.B.Schneider, M.Emmer, R.L.Perlman and I.Pastan, J.Biol.Chem. 246 (1971) 5929.

4. G.Zubay, D.Schwartz and J.Beckwith, Proc.Nat.Acad. Sci.U.S., 66 (1970) 104.

5. V.G.Allfrey, in: Histones and Nucleohistones, ed. D.M.P.Phillips (Plenum Publishing Co., London, 1971) p.241.

6. A.J.MacGillivray, J.Paul and G.Threlfall, Adv.Cancer Res., 15 (1972) 93.

7. G.Stein and R.Baserga, Adv.Cancer Res., 15 (1972) 287.

8. S.C.R.Elgin, S.C.Froehner, J.E.Smart and J.Bonner, Adv.Cell Mol.Biol., 1 (1971) 1.

9. A.E.Mirsky and H.Ris, J.Gen.Physiol., 34 (1951) 475.

10. C.W.Dingman and M.B.Sporn, J.Biol.Chem., 239 (1964) 3483.

11. J.H.Frenster, Nature, 206 (1965) 680.

12. F.Dolbeare and H.Koenig, Proc.Soc.Exptl.Biol.Med., 135 (1970) 636.

13. P.J.Helmsing and H.Berendes, J.Cell Biol., 50 (1971) 893.

14. K.Marushige and H.Ozaki, Devel.Biol., 16 (1967) 474.

15. R.J.Hill, D.L.Poccia and P.Doty, J.Mol.Biol., 61 (1971) 445.

16. G.R.Reek, R.T.Simpson and H.A.Sober, Proc.Nat.Acad. Sci.U.S., 69 (1972) 2317.

17. C.T.Teng, C.S.Teng and V.G.Allfrey, Biochem.Biophys. Res.Commun., 41 (1970) 690.

18. C.S.Teng, C.T.Teng and V.G.Allfrey, J.Biol.Chem., 246 (1971) 3597.

19. R.D.Platz, V.M.Kish and L.J.Kleinsmith, FEBS Lettr. 12 (1970) 38.

20. J.E.Loeb and C.Creuzet, Bull.Soc.Chim.Biol., 52 (1970) 1007.

21. F.Chytil and T.C.Spelsberg, Nature New Biol., 233 (1971) 215.

22. W.M.LeStourgeon and H.P.Rusch, Science, 174 (1971) 1233.

23. E.L.Gershey and L.J.Kleinsmith, Biochim.Biophys.Acta, 194 (1969) 519.

24. K.R.Shelton and J.M.Neelin, Biochemistry, 10 (1971) 2342.

25. G.Vidali, L.C.Boffa, V.C.Littau, K.M.Allfrey and V.G. Allfrey, J.Biol.Chem., in press.
26. K.R.Shelton and V.G.Allfrey, Nature , 228 (1970)132.
27. C.S.Teng and T.H.Hamilton, Biochem.Biophys.Res. Commun., 40 (1970) 1231.
28. V.Enea and V.G.Allfrey, Nature New Biol., in press.
29. K.Ahmed, Biochim.Biophys.Acta, 243 (1971) 38.
30. W.B.Benjamin and R.M.Goodman, Science , 166 (1969) 629.
31. V.G.Allfrey, C.S.Teng and C.T.Teng, in: Nucleic Acid - Protein Interactions - Nucleic Acid Synthesis in Viral Infection , ed., D.W.Ribbons, J.F.Woessner and J. Schultz (North Holland Publishing Co., Amsterdam (1971) p. 144.
32. C.S.Teng, C.T.Teng and V.G.Allfrey, Arch. Biochem. Biophys., in press.
33. E.M.Johnson and V.G.Allfrey, Arch.Biochem.Biophys. 152 (1972) 786.
34. R.W.Ruddon and C.N.Rainey, Biochem.Biophys. Res. Commun., 40 (1970) 152.
35. G.S.Stein and R.Baserga, J.Biol.Chem., 245 (1970) 6097.
36. G.Rovera and R.Baserga, J.Cell Physiol., 77 (1971) 201.
37. L.J.Kleinsmith, V.G.Allfrey and A.E.Mirsky, Science, 154 (1966) 780.
38. L.J.Kleinsmith, J.Heidema and A.Carroll, Nature, 226 (1970) 1025.
39. T.Y.Wang, Exp. Cell Res., 61 (1970) 455.
40. M.Kamiyama and T.Y.Wang, Biochim.Biophys.Acta, 228 (1971) 563.
41. D.Rickwood, G.Threlfall, A.J.MacGillivray and J.Paul, Biochem.J. , 129 (1972) 50p.
42. L.J.Kleinsmith, manuscript in preparation.
43. J.Paul and R.S.Gilmour, J.Mol.Biol., 34 (1968) 305.
44. R.S.Gilmour and J.Paul, J.Mol.Biol., 40 (1969) 137.
45. T.C.Spelsberg, L.S.Hnilica nd A.T.Ansevin, Biochim. Biophys.Acta , 228 (1971) 550.
46. G.Stein, S.Chaudhuri and R.Baserga, J.Biol.Chem., 247 (1972) 3918.

47. J.Farber, G.S.Stein and R.Baserga, Biochem.Biophys. Res.Commun. , 47 (1972) 790.
48. G.Stein and J.Farber, Proc.Nat.Acad.Sci.U.S., 69 (1972) 2918.
49. N.C.Kostraba and T.Y.Wang, Cancer Res., 32 (1972) 2348.
50. J.N.Davidson, S.C.Frazer and H.C.Hutchinson, Biochem.J., 49 (1951) 311.
51. R.N.Johnson and S.Albert, J.Biol.Chem., 200 (1953) 335.
52. T.A.Iangan, in : Regulation of Nucleic Acid and Protein Biosynthesis , ed. V.V.Koningsberger and L.Bosch, (Elsevier, Amsterdam, 1967) p.223.
53. L.J.Kleinsmith and V.G.Allfrey and A.E.Mirsky, Proc. Nat.Acad.Sci.U.S., 55 (1966) 1182.
54. L.J.Kleinsmith and V.G.Allfrey, Biochim.Biophys.Acta, 175 (1969) 123.
55. L.J.Kleinsmith and V.G.Allfrey, Biochim.Biophys.Acta, 175 (1969) 136.
56. W.B.Benjamin and A.Gellhorn, Proc.Nat.Acad.Sci.U. S. , 54 (1968) 262.
57. M.Kamiyama, B.Dastugue and J.Kruh, Biochem.Biophys. Res.Commun., 44 (1971) 1345.
58. M.Takeda, H.Yamamura and Y.Ohga, Biochem.Biophys. Res.Commun., 42 (1971) 103.
59. R.W.Ruddon and S.C.Anderson, Biochem.Biophys.Res. Commun. , 46 (1972) 1499.
60. M.Kamiyama, B.Dastugue, N.Defer and J.Kruh, Biochim. Biophys.Acta , 277 (1972) 576.
61. V.G.Allfrey, Cancer Res., 26 (1966) 2026.
62. V.G.Allfrey, Fed. Proc., 29 (1970) 1447.
63. T.A.Borgese and J.F.Bertles, Science , 148 (1965) 509.
64. F.L.Yu and P.Feigelson, Biochem.Biophys.Res.Com- mun., 35 (1969) 499.
65. W.Schmid and C.E.Sekeris, FEBS Lettr., 26 (1972)109.
66. L.D.Murthy, D.S.Pradhan and A.Sreenivasan, Biochim. Biophys.Acta , 199 (1970) 500.
67. T.A.Langan, Proc.Nat.Acad.Sci.U.S., 64 (1969)1276.

68. T.A.Langan, J.Biol.Chem., 244 (1969) 5763.
69. D.Yeung and I.T.Oliver, Biochemistry,7 (1968) 3231.
70. W.D.Wicks, F.T.Kenney and K.L.Lee, J.Biol.Chem., 244 (1969) 6008.
71. O.Greengard, Biochem.J., 115 (1969) 19.
72. J.-P.Jost, A.Hsie, S.D.Hughes and L.Ryan, J.Biol.Chem. 245 (1970) 351.
73. L.A.Dokas and L.J.Kleinsmith, Science, 172 (1971)1237.
74. P.B.Kaplowitz, R.D.Platz and L.J.Kleinsmith, Biochim. Biophys.Acta., 229 (1971) 739.
75. E.Miyamoto, G.L.Petzold, J.S.Harris and P.Greengard, Biochem.Biophys.Res.Commun., 44 (1971) 305.
76. E.M.Johnson, G.Vidali, V.C.Littau and V.G.Allfrey, J.Biol.Chem., in press.
77. J.Paul and I.R.More, Nature New Biol., 239 (1972)134.
78. K.Marushige and J.Bonner, J.Mol.Biol., 15 (1966) 160.
79. T.Y.Wang, Exp.Cell Res., 53 (1968) 288.
80. T.C.Spelsberg and L.S.Hnilica, Biochim.Biophys.Acta, 195 (1969) 63.
81. C.S.Teng and T.H.Hamilton, Proc.Nat.Acad.Sci.U.S., 63 (1969) 465.
82. M.Kamiyama, B.Dastugue and J.Kruh, Biochem.Biophys. Res.Commun., 44 (1971) 1345.
83. G.Vidali, L.C.Boffa and V.G.Allfrey, J.Biol.Chem., 247 (1972) 7365.
84. T.W.Borun and G.S.Stein, J.Cell Biol., 52 (1972) 308.
85. G.S.Stein and R.Baserga, J.Biol.Chem., 245 (1970)6097.
86. J.A.Smith, R.J.Martin, R.J.King and M.Vertes, Biochem. J., 119 (1970) 773.
87. R.Stellwagen and R.Cole, J.Biol.Chem., 244 (1969)4878.
88. D.Bootsma, L.Bidke and O.Vos, Exp.Cell Res., 33, (1964) 301.
89. G.S.Stein and T.W.Borun, J.Cell Biol., 52 (1972)242.
90. R.Hancock, J.Mol.Biol., 40 (1969) 457.
91. J.S.Bhorjee and T.Pederson, Proc.Nat.Acad.Sci.U.S., 69 (1972) 3345.
92. C.E.Zeilig, R.A.Johnson, D.L.Friedman and E.W.Sutherland, J.Cell Biol., 55 (1972) 296a.
93. A.Inoue and V.G.Allfrey, manuscript in preparation.

DISCUSSION

T.A. LANGAN: I wonder if the discrepancy that you observed between the action of cyclic AMP in vitro and in vivo might be due to the release of corticosteroid in the whole animal by injection of cyclic AMP and, with that question in mind, whether you have studied this in adrenalectomized rats.

V.G. ALLFREY: I should have specified that all the experiments using cyclic AMP were performed on adrenalectomized rats.

A.G. GORNALL: I would like to report some observations that Dr. C.C. Liew and a graduate student, D. Suria, have made. Dr. Liew developed his interest in this field while in Dr. Allfrey's laboratory and essentially the same method was used to fractionate the nuclear acidic proteins. Nuclei were obtained from kidneys of control and aldosterone treated adrenalectomized rats, 40 and 150 minutes after hormone injection. The acidic proteins were fractionated by electrophoresis in SDS-polyacrylamide gel. Incorporated ^{3}H-acetate and ^{32}P were measured after cutting the gel into 1 mm slices. Of the multiple protein bands four showed significant changes. The first showed acetylation stimulated by aldosterone at 40 minutes and still increased at 150 minutes. The second band showed increased acetylation at 40 minutes but had returned to normal at 150 minutes. A third band showed phosphorylation increased by aldosterone at 40 minutes with return to normal at 150 minutes. A fourth band was dephosphorylated at 40 minutes and was still depressed at 150 minutes. These effects, which appear to be selective, occurred during the so-called latent period of hormone action in the kidney. When liver nuclear acidic proteins were studied no such effects were seen, but this organ appears to respond to glucocorticoids rather than to mineralocorticoids.

G.H. DIXON: Do you have any comments, Dr. Allfrey?

V.G. ALLFREY: No, I am very happy to see these results.

C. ABELL: Have you studied the phosphorylated proteins that are formed following PHA stimulation of lymphocytes?

V.G. ALLFREY: I have not yet made a detailed study of the
proteins which appear after PHA stimulation of lymphocytes.
There have been studies of changes in nuclear acidic proteins in PHA-stimulated lymphocytes carried out by Drs.
Levy, Rosenberg and Simpson (reported in Biochemistry, 12,
224 (1973)), and they reported preferential synthesis of
several specific protein fractions shortly after exposure
to phytohemagglutinin.

C. ABELL: Are the membrane bound proteins phosphorylated
in those cells?

V.G. ALLFREY: In the cases that I have described, the
proteins are nuclear proteins rather than membrane contam-
inants. I might add that in the red cell experiments, we
were very concerned that some of the differences observed
might have been due to contaminating plasma membranes or
nuclear membrane constituents. We investigated this by
extracting the phenol-soluble proteins from isolated membrane fractions and displaying the proteins in the extract
on SDS-polyacrylamide gels to see whether they would constitute a serious threat to our interpretation of these
changes during development. The total contamination by
membrane proteins is very small, and the banding patterns
are easily distinguishable from those of the nuclear extract. Moreover, we have taken precautions to eliminate
plasma and cytoplasmic membrane contamination by purification of the nuclear fraction through sucrose density-grad-
ient centrifugation.
     Dr. C.C. Liew, who is now working in Dr. A. Gornall's
laboratory in Toronto, has also looked for contamination
of liver nuclear proteins by proteins derived from liver
nuclear membranes, and he concludes that no more than 8%
of the protein extracted in phenol from intact liver nuclei
could have been derived from the nuclear membranes. This
is in accord with observations made by Dr. Teng that the
polyacrylamide gel patterns are virtually identical for
phenol-soluble proteins prepared form isolated liver nuclei
and from liver chromatin fractions.

K.S. McCARTY: It is my impression from your studies of
HeLa synchronized cells that unlike the other systems that
you presented, there were minimal changes in the phosphorylation of acidic chromosomal proteins during the S-phase

and during mitosis.

V.G. ALLFREY: Phosphorylation peaks in the S-phase and in the early G1 phase of the next cell cycle.

K.S. McCARTY: Does this then represent a unique system? Do you have any speculations?

V.G. ALLFREY: Well, in the normal cell cycle of slime molds (that is when they are not prompted to undergo differentiation) there is very little change in the nuclear phosphoprotein pattern, as was shown for Physarum by LeStourgeon and Rusch. Changes in nuclear proteins do occur when these cells are induced to differentiate by changing the culture medium. Some protein bands virtually disappear while others become more intense. In HeLa cells there are small changes in the nuclear protein complement during the cell cycle, as reported by Drs. Bhorjee and Pederson (Proc. Nat. Acad. Sci. U.S. 69, 3345 (1972), and there are major differences in the rates of synthesis of nuclear acidic proteins during the cell cycle, as reported by Drs. Borun and Stein (J. Cell Biol., 52, 308 (1972).

K. SHELTON: I think you mentioned that the deacetylase was only in lymphocytes?

V.G. ALLFREY: No, the deacetylase is not limited to lymphocytes. It has been found in liver and in other tissues examined by Dr. Vidali. We did concentrate on the thymus lymphocyte because it provides an opportunity for the isolation of nuclei in non-aqueous media, and we could thus prove that the deacetylase is localized in the cell nucleus.

P.A. GALAND: I wondered if in a pure cell population, like HeLa cells you have at the same time changes in acetylation and phosphorylation, because the results on kidney could in fact deal with several types of cells, but in a pure population do both changes occur in the same moment in the same cell?

V.G. ALLFREY: I was not expecting to talk about acetylation. The acetylation of histones, which is very pronounced in histone fractions F2A1 and F3, is observed

247

in a number of cell types, regardless of the nature of stimulation employed. It is augmented in liver cells stimulated by cortisol, in lymphocytes stimulated by phytohemagglutin, and in kidney cells stimulated by aldosterone, as shown by Dr. Liew. Since in all these cases the acetylation of histones precedes the increase in rate of RNA synthesis, I was strongly convinced that genes could not be switched on unless the interaction between histones and DNA was weakened by acetylation. I still feel that the acetylation of histones does represent the mechanism for changing the binding of histones to DNA and that this is often important in starting new patterns of transcription. However, in the case of lymphocytes there are experiments by Dr. Ono and his collaborators in which the cells were stimulated by PHA, but the synthesis of new RNA's was blocked by the addition of cortisol to the culture medium. Under these conditions, histone acetylation is increased, the DNA binds more acridine orange and ^{3}H-actinomycin D, and it is clear that the physical state of the chromatin has been altered - but RNA synthesis is not turned on. It follows that a change in state is not, in itself, a sufficient condition for inducing the synthesis of new RNA's.

G.H. DIXON: Would you think it a necessary condition that you have to go through essentially two gates? You've got to change the state of the chromatin by perhaps changing the state of binding of the histones so that you would be reversing a sort of general negative control, but then you have to have some sort of positive effector which is going to allow RNA transcription from a specific segment of the genome.

V.G. ALLFREY: That is a mechanism I would like to consider for the activation of genes that exist largely in their repressed state within a cell nucleus, but I can also allow for operation of other, more direct systems of repression, such as the lac repressor in E. coli, where there is specific inhibition of transcription of particular gene loci. There are as well, other proteins that specifically initiate RNA synthesis, such as the cAMP-receptor protein, that operate without the involvement of histones.

C. ZEILIG: Approximately how many hours after release from the second thymidine block did you observe that first peak in phosphorylation of the acidic proteins?

V.G. ALLFREY: I think that the first peak is observed at three hours.

C. ZEILIG: And approximately how many hours into G1 did you see the second peak?

V.G. ALLFREY: It occurs early in G1. At that time, we are again dealing with a mixed, slightly asynchronous population in the early second cycle, but the peak probably occurs at just an hour or two into the G1 phase.

C. ZEILIG: We have recently reported on measurements of cyclic AMP levels in HeLa cells during the cell cycle and find that the levels begin to decline during G2 and reach a minimum corresponding to the peak of mitosis. And immediately 2 hours into G1 there appears to be another sharp peak or elevation in cyclic AMP.

V.G. ALLFREY: I am glad that you referred to this change in cyclic AMP levels, which I forgot to mention in my talk. It is referred to in the manuscript.

A. BRAUNWALDER: Can you influence the deacetylase activity for example by phosphoprotein phosphatase or any other kind of phosphatase?

V.G. ALLFREY: I know what you mean but I can not answer the question of what the effect would be of removing the phosphate groups from the histone deacetylase. It has not yet been tried. I believe that this phenomenon of histone deacetylation being controlled by a phosphoprotein and phosphoprotein phosphorylation being selectively influenced by different histones may be an important aspect of interlocking controls.

THE ROLE OF HISTONE PHOSPHORYLATION IN CELL DIVISION

ROGER CHALKLEY, ROD BALHORN, DENIS OLIVER and DARYL GRANNER
Department of Biochemistry
University of Iowa School of Medicine
Iowa City, Iowa 52242

Abstract: An analysis of a variety of cells in the dividing or in the non-dividing state has led to a substan-
tiation of the idea that histone phosphorylation is
invariably observed in dividing cells. The extent of
histone phosphorylation has been studied and we find
that essentially all the lysine-rich (F_1) histone
molecules are phosphorylated in a single, unique event
each cell cycle. Both newly synthesized and pre-
existing F_1 histones are phosphorylated in an event
which is temporarily associated with DNA synthesis,
though inhibition of DNA synthesis does not result in
an immediate inhibition of F_1 phosphorylation. Finally
it appears that histone phosphorylation and DNA synthesis take place at different sites within the nucleus.

It is well established that almost all cells contain
five major groups of histone in a tight association with
the chromosomal DNA (1,2). Histones are highly basic proteins and much of the interaction between histone and DNA
is electrostatic in nature though a contribution from
hydrophobic forces is also recognized (3). Knowledge of
the chemistry of the five major histone fractions has
become increasingly precise during the last five years and
the sequences of four of the histones have been determined,
and about half of the sequence of the largest remaining
histone is known. In general, the primary sequence of
histones shows an astonishing conservatism over vast
periods of evolutionary history (4). All the available

251

evidence indicates that the four histone fractions, F_{2a1}, F_{2a2}, F_{2b} and F_3, have had the lowest rate of evolutionary sequence changes of any proteins thus far studied. This is seen most dramatically for the arginine-rich histone fractions, F_{2a1} and F_3, which have shown only two conservative amino acid replacements (4) during the divergence of peas and cows (for F_{2a1}) or which demonstrate only one amino acid replacement in the time period between the evolution of fish and mammals (F_3). In contrast to the four histone fractions described above, the lysine-rich (F_1) histone appears to undergo changes in amino acid sequence (5) and molecular weight in different creatures at a much higher rate. In fact it will not be surprising if this histone turns out to have one of the higher mutational rates among the various proteins studied.

The rapid advances in our knowledge of the chemistry of histones have influenced much of the thinking on the nature of their biological role. Initially it was thought that there were a large number of different molecules of histone and that they might play a specific role in gene control in differentiated organisms (6). However, the notion that histones were highly heterogeneous derived its force from preparations which were grossly impure and degraded, and as it has become established that the heterogeneity of histones is limited, so has the concept that histones are involved in specific gene repression receded. It now seems clear that since most eucaryote cells contain essentially the same histone molecules (except for F_1) then the histones of the pea plant or the fish must be doing the same thing as the histones in the mammalian nucleus. A much favored idea at this time is that histones are involved in recognizing those features of DNA which are in common in all organisms and that they subsequently facilitate the ability of DNA to adopt a compact conformation necessary to permit both its existence in the confined space of the nucleus, and its subsequent collapse into the compact form of the mitotic chromosomes. So far as we know these two requirements are held in common by the DNA of all eucaryotes, and as such, might yield a function for a group of protein associated with DNA, which might permit of little sequence change during evolution, since the overall structure of DNA (linear, rod-like, polyanion) has likewise remained unchanged.

Recent advances in the physical chemistry of isolated chromosomal material using X-ray (7) and CD spectroscopic methods (8) together with hydrodynamic studies (9) have provided convincing evidence that histones can compel DNA to adopt a much more compact conformation. This, of course, does not exclude histones from exerting an influence over genetic transcription as presumably control over the tertiary structure of DNA could be used to control the ability of RNA polymerase to bind to the appropriate part of the chromosome. Again one runs into the problem of how histones with their limited heterogeneity could exert control in a specific manner. Most workers in this area have recently tended to concentrate on an involvement of other factors such as RNA or non-histone nuclear proteins which might act in concert with the histones to exert a control over gene readout (10,11). At this time, however, there is no well documented evidence that histones have any control over gene activity.

There has been considerable controversy concerning the metabolic stability of histones. There is, however, a substantial body of opinion which argues that all histone fractions are metabolically stable (12-14) (i.e., that they do not turnover at a significant rate and thus exist as long as the DNA with which they combine).

Thus, one might be tempted by the notion that since histones are unusually inert and passive, they function simply to modify the conformation of DNA. However, although the histone molecules remain intact, there is a substantial metabolic activity involving the histone molecules. Apparently in all cells, except mature spermatazoa, histones F_{2a1} and F_3 are acetylated at specific lysine residues (15). The acetate groups turnover at a moderate rate. The removal of a positive charge due to acetylation has rendered this type of molecule amenable to electrophoretic separation from the parent unacetylated molecule, thus giving rise to an electrophoretic heterogeneity in which all the components have the same primary amino acid sequence. The function of the repeated acetylation remains a mystery, though Allfrey and his co-workers (16) have reported a correlation between the extent of acetylation of these histones and the genetic activity of the cell.

Electrophoretic heterogeneity is also observed within the lysine-rich histone fraction (F_1). This is due in part to an heterogeneity in sequence and molecular weight (5). This histone fraction consists of between 3-5 molecules, which, while preserving the general features of the lysine-rich histone, nonetheless show small differences in sequence which can be detected electrophoretically (17). On occasion additional electrophoretic heterogeneity is introduced into this histone as a result of the addition of one or more phosphate groups to specific serine residues within the various F_1 molecules (18). In contrast to the acetylation of the more arginine-rich histones, bulk phosphorylation does not occur in all cells but as we will see below is restricted to dividing cells, which, of course, has led to the proposal that lysine-rich histone phosphorylation has an indispensable role to play in the overall process of cell division. The nature of this process and its control form a part of this presentation and are the subject of active research in several laboratories. In contrast to the division-associated phosphorylation in which probably all lysine-rich histones are phosphorylated, a low level of histone phosphorylation at a separate site has been described by Langan (19). It is important to distinguish between the two types of phosphorylation as it is likely they have separate and distinct biological functions.

Association of F_1-phosphorylation with cell division

Earlier work of Ord and Stocken had indicated that F_1 phosphorylation was elevated in dividing cells by a factor of some 2-3 fold over that in non-dividing cells (20). It later became clear that the relative level of increase in phosphorylation had been grossly underestimated due to a massive contamination of isolated lysine-rich histone by inorganic phosphate and by small polyribonucleotides. This is shown in Table 1, in which results are presented of an experiment in which ^{32}P was injected into sham operated and partially hepatectomized rats some 30 hours after the hepatectomy. Lysine-rich histone was isolated by the method used by the previous workers and its specific activity determined. Since we suspected that anionic impurities might be present, the histone fractions were purified by column chromatography on a cation exchange resin. It is clear that both sham operated and regenerating liver F_1

isolated in the standard manner are grossly contaminated with ^{32}P which is not bound to histone. It is also seen that the relative increase of ^{32}P found in the dividing cells is proportionally much greater when impurities are removed.

TABLE 1

Specific activity of ^{32}P-labelled F_1 histone (cpm/mg)

Unpurified sham F_1	39,600
Column purified sham F_1	1,640
Unpurified regenerating liver F_1	140,000
Column purified regenerating liver F_1	11,360

Recent experiments have shown that a single passage through the resin does not remove all the inorganic phosphate which binds histone tenaciously, but that further purification removes almost all the ^{32}P associated with the sham operated liver F_1 histone, so that the level of increase in phosphorylation in regenerating liver F_1 histone approaches 50-fold.

Such purification schemes require large quantities of F_1 histone, and are moreover lengthy and tedious. Accordingly, before beginning a detailed analysis of a correlation between cell division and histone phosphorylation, we found it necessary to establish appropriate criteria for a rapid, but precise, estimate of the occurrence and the extent of phosphorylation of this histone fraction.

Criteria for assay of histone phosphorylation
The electrophoretic system we employ operates at pH 2.8. At this pH, the introduction of phosphate groups to a highly positively charged protein reduces its electrophoretic mobility by slightly more than 1% for each phosphate group. Thus a series of equispaced electrophoretic bands is indicative of the presence of phosphate groups. However, additional documentation is required since acetate-

induced microheterogeneity is well known, and also because the lysine-rich histone is inherently heterogeneous as discussed above. The criteria we apply for analysis of histone F_1 phosphorylation are: (1) the presence of electrophoretic heterogeneity; (2) following an incubation of ^{32}P with the intact cells, the isolated F_1 histone should show the presence of ^{32}P associated with the more slowly migrating electrophoretic bands. The specific activity of the bands should increase inversely as the rate of migration; (3) incubation of the F_1 histone with alkaline phosphatase should remove not only the associated radiolabel (^{32}P), but also should abolish the phosphate-induced microheterogeneity. This provides a means for specifically assaying for phosphorylation in a system in which it is difficult to perform the ^{32}P labelling studies.

Tests for division-associated phosphorylation

Armed with these criteria it was possible to test the notion that F_1 phosphorylation was specifically associated with cell division. We have utilized several systems comparing a dividing cell with a similar type of cell but which is not dividing. The systems studied have been: (1) normal rat liver compared to regenerating rat liver; (2) normal tissues compared to tumors derived from these tissues; (3) normal (adult) rat liver compared to fetal rat liver in which essentially all cells are dividing; and (4) normal rat liver compared to a tissue culture line derived from rat liver. The latter type of cells were analyzed both in exponential growth phase and in stationary phase.

The electrophoretic patterns of rat liver lysine-rich histone 38 hours after partial hepatectomy are shown in Fig. 1. It is seen that there is increased heterogeneity in the regenerating liver F_1 histone relative to the sham operated liver F_1. The new bands are more slowly moving than the parent F_1 bands. Furthermore, pre-treatment with alkaline phosphatase abolished the band heterogeneity induced by partial hepatectomy, and the F_1 pattern now has a pattern indistinguishable from adult rat liver F_1. There is a measure of heterogeneity in the adult rat liver F_1 histone but this is not affected by phosphatase treatment, and we suspect it is a reflection of sequence microheterogeneity as discussed previously.

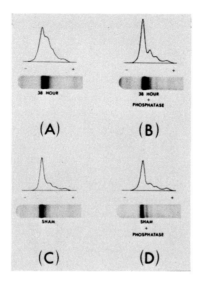

Fig. 1. Phosphatase treatment of lysine-rich histones. (A) Untreated 38 hr regenerating liver; (B) phosphatase-treated 38 hr regenerating liver; (C) untreated sham operated liver F_1; and (D) phosphatase-treated, sham operated liver F_1.

The gels shown in Fig. 1 were sectioned and counted to analyze for band-associated ^{32}P. The results of such an approach are shown in Table 2, in which we document the specific activity of bands in the F_1 region.

TABLE 2

^{32}P incorporation into F_1 histone subfractions

Hr after partial hepatectomy	Specific activity (cpm/mg)			
	a	b	c	d
Sham	75	200	230	n
29	12	580	2100	4275
29 + phosphatase	n	n	n	n

n = not detectable.

257

Band (a) is the most rapidly migrating and (d) is that most slowly moving in the electrophoretic field. Three points emerge. (1) The more slowly moving bands have the highest specific activities, indicating multiple phosphorylation of a single parent molecule; (2) phosphatase treatment has effectively removed the associated label; and finally (3) the level of phosphorylation of the F_1 histone is much greater after the induction of cell division by the hepatectomy operation.

After partial hepatectomy it is known that the subsequent waves of DNA synthesis and cell division are partly synchronized, though the precise timing depends on the strain of rat used and on other variables. Nonetheless, we wondered if the extent of histone phosphorylation would vary as a function of time after the operation. That this is indeed so is shown in Fig. 2. Two peaks of phosphorylation are noted, at the 28th and the 40th hour. The preci-

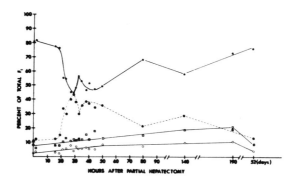

Fig. 2. Changes in the lysine-rich histone subfractions during rat liver regeneration. Subfractions are: *, most rapidly migrating (parent) band; ●, monophosphorylated, second most rapidly migrating form; ■ , third electrophoretic sub-band. Each point is the mean of 3 separate operations and histone isolations.

sion of the synchronized bursts of DNA synthesis and mitosis does not permit us to correlate the observed phosphorylation specifically with either of these events. Though it is noteworthy that histone phosphorylation is proceeding apace during the 18th-22nd hour which is normally thought of as a time of DNA synthesis initiation.

The developing, fetal rat liver provides an attractive model for a dividing liver cell system. The cell generation time is approximately 14 hrs, and all cells are dividing in the 15 day fetal liver. Immediately before birth (day 20) all cells are still involved in division, but the generation time has increased to 22 hours (21). After birth the number of cells involved in division falls to 5% at the 56th day and the time of G_1 and S both increase with age. Lysine-rich histone was extracted from fetal and young rat livers, and analyzed for electrophoretic microheterogeneity. The results are shown in Fig. 3.

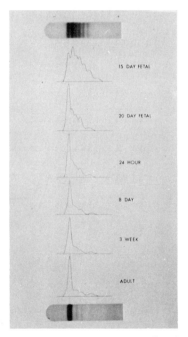

Fig. 3. Changes in phosphorylation-induced microheterogeneity of the lysine-rich histones of rat liver at different stages in development.

Clearly the amount of heterogeneity in the 15 day fetal rat exceeds anything we have seen previously. However, this is also the most rapidly dividing system we have used thus far. The fetal tissue, in which all cells are dividing and the S phase occupies much of the cell cycle, shows massive phosphorylation. At day 20, even though all cells are still dividing, the S phase now occupies a smaller fraction of the cell cycle and the level of phosphorylation falls. As the rat deveops the cell cycle lengthens and the fraction of cells involved in division falls, then the level of phosphorylation falls until it is essentially non-existent some 8 weeks after birth.

We have recently completed an analysis of a wide variety of tumor histones (22), concentrating mostly on the heterogeneity of the lysine-rich histone. Again we find a precise correlation between the extent of phosphatase-sensitive microheterogeneity and the rate of cell division. This is most easily seen in a series of Morris hepatomas of different growth rates as shown in Fig. 4, though we emphasize that this represents results typical of both rat and mice tumors, and from many tissues of origin other than liver.

The work with the tumor systems led us to the conclusions that: (a) lysine-rich histone phosphorylation is increased in rapidly growing tumors; (b) this event is not merely a liver- or hepatoma-associated phenomenon; (c) F_1 phosphorylation is not rat tissue specific but has been found in all mouse tumors as well, thus supporting the proposal that it is a more general phenomenon directly associated with cell replication; and (d) electrophoretic differences in the histone complement of normal and tumor cells lie only in the lysine-rich histone phosphorylation-induced electrophoretic microheterogeneity found in dividing cells.

Cultured cells provide an attractive system for studying histone phosphorylation. Not only can the experimental procedure be more precise than when using whole animals (particularly for pulse-chase studies); but also the cells will proliferate exponentially when subcultured into fresh medium progressing to stationary phase at higher cell densities. In stationary phase the rate of DNA synthesis is essentially indetectable and thus one can employ a single cell line to analyze for histone phosphorylation in either dividing or non-dividing cells.

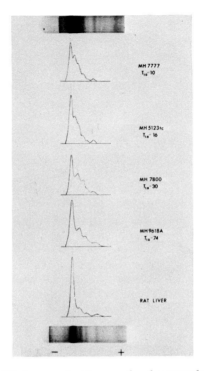

Fig. 4. High-resolution gel electrophoretic patterns and microdensitometer scans of lysine-rich histones from the Morris hepatomas (MH). $T_{1.0}$ is the time
in days for the tumor to reach a wet weight of 1.0 g
following a standard innoculum.

The high resolution electrophoretic patterns of the
lysine-rich histones from exponentially growing HTC cells,
and stationary phase HTC cells are presented in Fig. 5
along with a sample of exponential phase F_1 treated with
alkaline phosphatase. As is customary, the presence of
phosphorylation in exponentially growing cells has led to
an accumulation of slower migrating F_1 bands. So much so,
that the second fastest band is now present in the largest
amount. Other yet more slower moving bands are distinguishable indicating several levels of phosphorylation.
Phosphatase treatment reduces the heterogeneity to a major
and a minor band (presumably reflecting sequence microheterogeneity) which appear to be identical to the pattern
obtained from stationary phase cells.

261

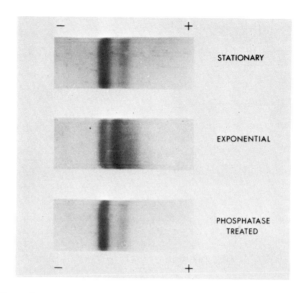

Fig. 5. High resolution electrophoretic analysis of F_1 from exponentially growing HTC cells and from stationary phase HTC cells. Also included is the electrophoretic pattern of exponentially growing HTC-F_1 after phosphatase treatment.

We conclude that the lysine-rich histone is not phosphorylated in stationary phase cells. Since these cells had been in stationary phase for approximately 24 hours, this implies a fairly rapid rate of F_1-phosphate turnover.

The rate of turnover of lysine-rich histone phosphate was studied after pulsing exponentially growing HTC cells with ^{32}P-phosphate for three hours. The cells were removed from the radioactive medium, washed and resuspended in fresh medium. At appropriate times thereafter cells were removed for analysis. The analysis consisted of isolation of the F_1 histone, electrophoretic separation, excision of the various electrophoretic sub-bands and counting associated radiolabel. In this way we can observe decay of ^{32}P in mono-, di- and multiply phosphorylated species. The results of such an experiment are shown in Fig. 6 and shows that ^{32}P label turns over with a half life of about 4.5 hours.

262

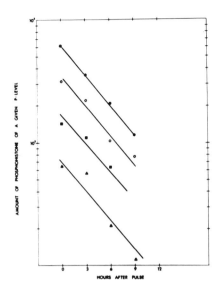

Fig. 6. Loss of ^{32}P-phosphate from various phosphorylated forms of HTC-F_1 histone as a function of time. ●-●, monophosphorylated F_1; o-o, diphosphorylated F_1; ■-■, triphosphorylated F_1 and ▲-▲, tetraphosphorylated F_1.

The kinetics of removal are pseudo–first order for all levels of phosphorylation. Furthermore, the observation that each level of phosphorylation has the same kinetics of hydrolysis would seem to indicate that a given lysine-rich histone loses all its associated phosphate groups at the same time. Most likely the rate-limiting step is the binding of the phosphatase and it subsequently removes all phosphate groups before being released. The observed half-life is consistent with the absence of detectable amounts of phosphorylated species in cells which had remained in stationary phase for 24 hours as described above.

Thus, utilizing a variety of different cell types we conclude that a substantial level of phosphorylation of the lysine-rich histone is a characteristic of dividing cells, and that moderately rapidly after the phosphorylation event the phosphate is removed.

Molecular nature of F_1-histone phosphorylation

Having established a correlation between cell replication and histone phosphorylation, we have more recently attempted to find out something about the biological function of histone phosphorylation, in the hope that it will throw some light on histone function itself. One route of attack has been to ask if pre-existing or newly synthesized histone were to be phosphorylated. For instance, if it proved to be the latter one might consider a possible role in terms of deposition on the chromosome. We need to know whether it is a unique event or if a given histone can be repeatedly phosphorylated throughout the division cycle. We have also asked questions concerning both a possible temporal and a spatial link to the act of DNA synthesis.

Phosphorylation of both newly synthesized and pre-existing histones

Since histones do not turnover appreciably, a newly synthesized chromosome should consist of a half complement of histone from previous rounds of chromosome replication (old histone) together with a half complement of newly synthesized (new) histone. We have devised an experiment to show whether new or old (or both) histones are phosphorylated. HTC cells were grown in the presence of ^{3}H-lysine for 3 hours, thus ensuring that histones synthesized in this time would be labelled. The cells were collected and washed. Half of the cells were resuspended in fresh medium and the remainder used for an immediate isolation of F_1 histone. After 24 hours of cell growth in fresh medium the label would now be in old histone (the cell generation time is 24 hours); the cells were collected and F_1 isolated. The F_1 histones from the two parts of the experiment were subjected to electrophoresis and the sub-bands analyzed for the presence of ^{3}H-lysine. In this way we could ask directly whether a histone containing ^{3}H-lysine had shifted to a phosphorylation band (i.e., had been phosphorylated). The result of such an analysis is shown in Fig. 7. It is seen that cells collected at 0 hr after the termination of the pulse are phosphorylated (i.e., new histones). The observation that 24 hours later the ^{3}H-containing histone (now old histone) is found in the phosphorylated bands provides convincing evidence that old histone is likewise phosphorylated. Unless the phosphorylation of new and old histone is unrelated (and we cannot exclude this), then

264

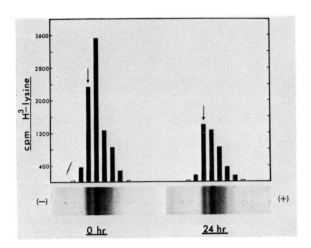

Fig. 7. Incorporation of ^{3}H-lysine into phosphorylated and parent lysine-rich histones. Cells were collected for F_1 isolation at 0 and 24 hrs after the completion of the 3 hr pulse of ^{3}H-lysine

this observation argues that phosphorylation is not a transportation related device required by newly synthesized histone.

F_1 phosphorylation is a single, unique event

We sought to determine whether the phosphorylation of an F_1 molecule is repetitive in nature (i.e., does the F_1 molecule become phosphorylated and dephosphorylated several times in the course of a cell cycle) or whether it is simply a single, unique event for each F_1 molecule within each cell cycle. In addition, we wanted to know whether there was a time lapse between synthesis of an F_1 histone and its subsequent phosphorylation. These points were probed by employing a very short (20 min) pulse of ^{3}H-lysine and analyzing for ^{3}H-lysine in phosphorylated F_1 bands as described above. After the pulse the cells were washed and resuspended in fresh medium and harvested at appropriate times into the chase.

The results from this experiment are shown in Fig. 8.

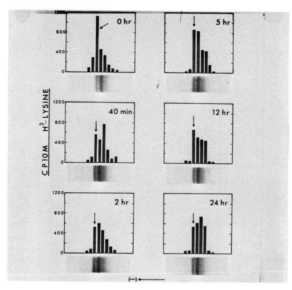

Fig. 8. Incorporation of a short (20 min) pulse of
³H-lysine into the various phosphorylated forms of F_1
during one cell cycle. The major parental histone band
is denoted by an arrow.

Immediately following the pulse (i.e., at zero time) most
of the radiolabel is associated with parental lysine-rich
histones. After a 40 min chase, a dramatic shift in the
electrophoretic position of the ³H-labelled histone can be
seen, so that the large bulk of the label is now associated
with those F_1 molecules which are phosphorylated. The observation of this lag period most likely excludes a function of phosphorylation as a transport device and argues
that the histone is associated with the chromosome before
it is phosphorylated. It also argues (for new histone at
least) that phosphorylation cannot be directly associated
with a mitotic event as it occurs too soon after synthesis.
As the chase is extended we observe that the radioisotope
in the lysine-rich histone shifts back towards the parent
unphosphorylated form, though as expected from the half-
life of dephosphorylation, the return to the parent form
is not complete. The half-life of ³H in the phosphorylated
bands is approximately 5 hours and thus agrees well with
the value observed for direct phosphate removal. This
demands that phosphate groups are not being added to his-

tone molecules repetitively during a single cell cycle (i.e. phosphorylation is a unique event). Though of course if the experiment is extended to 24 hours after the initial pulse, then once more the labelled histone reappears in the phosphorylated position which we interpret as phosphorylation of old histone as discussed above.

If one extrapolates the time course of phosphorylation back to zero time, it appears that greater than 85% of all newly synthesized F_1 histone is phosphorylated. Our inability to pulse label old histones, effectively precludes us testing if this is true for old histone. However, it is necessary to assume this point in order to account for the amounts of parent (30%) and phosphorylated F_1 (70%) formed in exponentially growing HTC cells, considering the observed rate of turnover of the phosphate groups.

The observation that most (or all) of the F_1 molecules synthesized in a short pulse are phosphorylated in a short time period, and not subsequently phosphorylated until the next cell cycle, suggests that they must be deposited upon the chromosome in a coherent and organized fashion. We suspect that histones are deposited sequentially on the chromosome and that several minutes later they are phosphorylated. Also, both new and old histones are phosphorylated and although this could be occurring at different sites it would appear to make for a more simple process if they were phosphorylated coincidentally. If this is so, we may expect that phosphorylation is post-replicative (i.e., occurs on newly synthesized DNA) unless deposition is pre-replicative, and recent evidence argues against this point and we favor the idea that histones are deposited upon the DNA immediately after the replication fork.

Dependence of histone phosphorylation upon DNA synthesis

Histones are synthesized during the S phase of the cell cycle, and since phosphorylation occurs within some 20 min after synthesis we expected that direct measurement of ^{32}P incorporation into the lysine-rich histone should occur in a temporal association with DNA synthesis. That this is indeed the case is shown in Fig. 9. HTC cells were synchronized by colcemid block in metaphase, after release of the mitotic block aliquots of the cells now traversing the cell cycle in synchrony, were incubated with either ^{32}P or

[3]H-thymidine to assay for F_1-phosphorylation and for rate of DNA synthesis respectively. The data of Fig. 9 show that phosphorylation of histone F_1 (and F_{2a2}) is at a maximum at the time that DNA synthesis has reached its greatest value. Essentially no phosphorylation of the remaining histone fractions was observed.

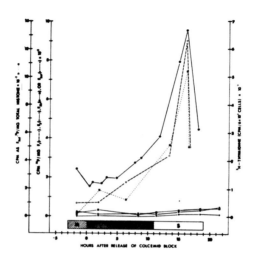

Fig. 9. Correlation of DNA synthesis and ^{32}P-phosphate incorporation synchronized HTC cells.

We wondered what effect the inhibition of DNA synthesis would have on histone phosphorylation. The inhibitors we have utilized have been cytosine-arabinoside (Ara-C), hydroxyurea and cycloheximide (which promptly inhibits DNA synthesis as a result of inhibition of protein (histone?) synthesis). The data of Fig. 10 show that the rate of phosphorylation is slowly reduced so that after 4-8 hrs in the presence of the inhibitor a basal level of some 40-50% of the initial rate is obtained. In general, the different inhibitors of DNA synthesis all exert a rather similar effect on histone phosphorylation. Clearly in the case of cycloheximide only pre-existing histone continues to be phosphorylated, but there is not likely to be sufficient to maintain this rate of phosphorylation for 8 hours in this population of S phase cells unless the event also occurs on chromosomes not immediately replicating when the

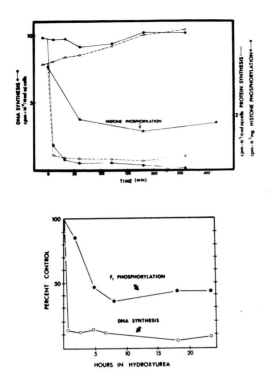

Fig. 10. (a) Effect of cycloheximide on DNA synthesis, histone synthesis and F_1-phosphorylation. (b) Effect of hydroxyurea on DNA synthesis and F_1-phosphorylation.

inhibitor was added. The other most likely explanation for these observations is that the phosphorylation event ceases to be unique (non-repetitious) in the presence of cycloheximide and that an equilibrium level is set up between phosphorylation and phosphate hydrolysis over a specific segment of the chromosome along which phosphorylation was occurring when the inhibitor was added. The inference from this interpretation is that the histone kinase is spatially localized within the nucleus.

In this cell line the concentration of hydroxyurea used has little effect on the rate of histone synthesis (in contrast to observations in HeLa cells). However, experiments have shown that within 4 hours after administration of the

269

drug, the newly synthesized histone is no longer phosphorylated, and yet phosphorylation of F_1 continues as shown in
Fig. 10, though at a reduced rate. Presumably the F_1 which
is being phosphorylated is at a different site and again we
suspect that the event may have ceased to be non-repetitious. This point is currently under active investigation;
however, we will not be surprised to find that the aspect
of phosphorylation most tied to the act of DNA replication
is the non-repetitious nature of the event.

Lack of spatial correlation between DNA synthesis and his-tone phosphorylation

There is a membrane fraction of isolated chromatin
which contains DNA of high specific activity after a brief
pulse of ^{32}P or ^3H-thymidine, but the specific activity
falls dramatically if the length of the pulse is extended.
We take this as a strong indication that this is a site of
DNA synthesis in vivo. The membrane fraction is probably
derived from the nuclear membrane. Be this as it may, it
provides us with an operational definition of a DNA synthetic site, and as such we can ask if the specific activity
of F_1-phosphate behaves in a similar manner to DNA after
pulses of different time periods. The data of Fig. 11
show that whereas the specific activity of DNA at the membrane site relative to that in the nucleoplasm, is 35 after
a 90 sec pulse and decreases to 2 after 30 minutes; the
relative specific activity of histone phosphate associated
with this site behaves in a contrary fashion, so that it
actually increases with increased pulse length.

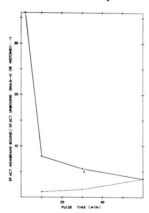

Fig. 11. Relative specific activity of membrane-
associated newly synthesized DNA (•——•) and newly phosphorylated F_1 histone (•--•) as a function of time of
pulse.

270

We conclude from this observation that DNA synthesis and histone phosphorylation take place at different sites and are separated by at least the length of the DNA obtained in these sheared fragments ($\sim 2.10^6$).

Histone phosphorylation, then, is an event which involves most, or all, lysine-rich histone molecules in the cell nucleus in what is most probably a highly organized sequential process occurring some 20 minutes after deposition of newly synthesized histone onto the replicating chromosome. We envisage that F_1 histone phosphorylation takes place post-replicatively (after the replication fork), though at this time this last point is perhaps no more than an enlightened speculation, depending strictly on where newly synthesized histones are deposited on the chromosome.

What do these observations and inferences tell us about the biological function of histone phosphorylation? Sad to relate, it must be confessed that it tells us little, though we are consoled by the knowledge that this area of histone research is still in its infancy. However, the lysine-rich histone is well known for its propensity to cross-link nucleohistone strands. It may be that the daughter strands after replication and histone deposition are cross-linked by the F_1 histones. The introduction of phosphate groups at critical points in the molecule might well act as a switch releasing the daughter chromosomes from their early embrace. Subsequent removal of the phosphate groups would permit the F_1 histones to become involved in appropriate cross-linking within the single chromosome of which they form a part. This model is presented largely as a contentious speculation, particularly bearing in mind that part of the charm of the histone field has been the great frequency with which rock-firm Shibboleths have collapsed into the dust of later studies.

Acknowledgment
    This work would not have been possible without the pioneering work of Drs. Sakol Panyim and David Sherod. We thank the U.S. Public Health Service for financial support through grants GM-46410 and CA-10871.

271

References

1. D. M. P. Philips and E. W. Johns, Biochem. J., 94 (1965) 127.
2. S. Panyim, D. Bilek and R. Chalkley, J. Biol. Chem., 246 (1971) 4206.
3. J. A. Bartley and R. Chalkley, J. Biol. Chem., 247 (1972) 3647.
4. R. J. Delange, D. M. Fambrough, E. L. Smith and J. Bonner, J. Biol. Chem. 244 (1969) 319.
5. S. C. Rall and R. D. Cole, J. Biol. Chem., 246 (1971) 7175.
6. R. C. C. Huang and J. Bonner, Proc. Natl. Acad. Sci., 48 (1962) 1216.
7. J. F. Pardon, M. H. F. Wilkins and B. M. Richards, Nature, 215 (1967) 508.
8. T. Y. Shih and G. D. Fasman, Biochemistry, 9 (1971) 2814.
9. J. A. Bartley and R. Chalkley, Biochim. Biophys. Acta, 160 (1968) 224.
10. C. S. Teng, C. T. Teng and V. G. Allfrey, J. Biol. Chem. 246 (1971) 3597.
11. W. Benjamin and R. M. Goodman, Science, 166 (1969) 629.
12. P. Byvoet, J. Mol. Biol., 17 (1966) 311.
13. R. Hancock, J. Mol. Biol. 40 (1969) 457.
14. R. Balhorn, D. Oliver, P. Hohmann, R. Chalkley and D. Granner, Biochemistry, 11 (1972) 3915.
15. V. G. Allfrey, R. Faulkner and A. E. Mirsky, Proc. Natl. Acad. Sci., 51 (1964) 786.
16. B. G. T. Pogo, V. G. Allfrey and A. E. Mirsky, Proc. Natl. Acad. Sci., 55 (1966) 805.
17. M. Bustin and R. D. Cole, J. Biol. Chem., 244 (1969) 5291.
18. R. Balhorn, W. O. Rieke and R. Chalkley, Biochemistry, 10 (1971) 3952.
19. T. A. Langan, S. C. Rall and R. D. Cole, J. Biol. Chem. 246 (1971) 1942.
20. M. G. Ord and L. A. Stocken, Biochem. J. 112 (1969) 81.
21. R. Balhorn, M. Balhorn and R. Chalkley, Devel. Biol., 29 (1972) 199.
22. R. Balhorn, M. Balhorn, H. P. Morris and R. Chalkley, Cancer Res., 32 (1972) 1775.

DISCUSSION

C. MOORE: What is the effect of phosphorylated histones on the phosphorylation on histones? Do you observe any product inhibition? I ask this question because you presented a slide showing that in the absence of DNA synthesis, the amount of phosphorylation of histone decreased to about 40% of the control value.

R. CHALKLEY: I would imagine that what we are seeing there, if I am right about the shift toward the repetitive phosphorylation in the presence of the drug is that one sets up an equilibrium between rate of phosphorylation and dephosphorylation. I have no experiments I can report on where we have assayed the effect of one phosphohistone on the phosphorylation of another histone.

C. MOORE: But, the phosphorylation of the histone can be done in vitro?

R. CHALKLEY: All of the experiments I have described have been done in vivo. The net phosphorylation we observed is probably the result of an equilibrium between phosphorylation and dephosphorylation. We are now setting up nuclear systems which will continue to make DNA and we are now assaying for histone phosphorylation of these systems, to see if it behaves in a similar fashion to the other experiments I have described.

R. SHARMA: It is my understanding that histones are very unstable even at 0° C? Yet during your electrophoretic analysis, which must have taken some time, you did not see any degradation of these histones?

R. CHALKLEY: The point upon which I did not make myself clear, is that it is during the isolation of histone, when you are still at the stage of chromatin, that there is a very active protease present. Once you treat with acid, and isolate the histones, and certainly once on the gel, in an acidic environment, they do not undergo any degradation.

K. SHELTON: I was very struck by your idea that the phosphorylation might permit the DNA strands to separate. But so that you did not have lysine-rich histone floating around in the nucleus, it would be nice to keep one end stuck. Do you suppose that, if you looked at the histone molecule, you might find phosphorylation restricted to either one or the other of the sticky basic ends?

R. CHALKLEY: I am not sure. I really do not envision histones floating around in the nucleus. I rather think of histone phosphorylation as a switch.

K. SHELTON: It could be interesting if either o ne end or the other would be phosphorylated.

R. CHALKLEY: I have a hunch that there is no division associated phosphorylation in the amino end, and I think from what Dr. Dixon has done, that we are probably going to see it somewhere near the middle.

C. MOORE: If indeed the histones are unstable in the cell, (and one of your experiments points to a relationship between the age of the preparation, the histone and the phosphorylation mechanism) and if you are working with lysine-rich histones which are being degraded, what happens to the lysine residues? Are they being turned over into new histones? In essence what is the true importance of the phosphorylation at which you are looking in terms of the fate of the lysine you are observing?

R. CHALKLEY: The evidence now, is that histones are very stable in the cell, perhaps not quite as stable as DNA but, at least precious near that, so it is hard to detect any turnover of histone. The point is that histone degradation is something that occurs once you start blending the tissue, and from that moment until you are into acid, then you have a danger. In fact, there are inhibitors we use to try and slow down or stop the proteolysis. But, this aspect of hisone instability is an artifact. It is a tremendous artifact, but it is simply unavoidable with some tissues.

G.H. DIXON: Could I just add a comment there? I would like to put in a commercial for trout testis chromatin. It is very, very stable. After many hours at 0° C, it does not seem to subject to appreciable proteolysis.

R. CHALKLEY: I think the worse chromatin that I know of
is calf thymus chromatin, and it is really tragic because
calf thymus is a good source of chromatin, it has a
lot of large nuclei, the contamination problem is minimal,
but the proteolysis problem is maximal in this system,
and it just happened to be the system most people used in
much of the early histone work.

. HIGLAND: Do you feel that the base composition of the
DNA predetermines the amount of phosphorylation of the
histone?

R. CHALKLEY: I guess the honest answer, is that I have no
idea. But, I think I could add that if you extrapolate
that time curve that I showed, showing the substantial
bulk of phosphorylation after a short time, in which we
estimate about 85% of all lysine-rich histones are phosphorylated. In fact, I would not be surprised at all,
if histone was phosphorylated once each time the cell di-
vides. Thus, one would not expect DNA composition to
effect histone phosphorylation.

L.R. GURLEY: I would like to comment on the question of
whether old f1 is phosphorylated prior to new f1. When
looking at histone phosphorylation in synchronized CHO cells
using the isoleucine deprivation method, we find the same
thing as you, that there is no phosphorylation of f1 in
G1 arrested cells. We have studied the kinetics of f1
phosphorylation in G1 as the cells approach S phase and
find that the phosphorylation of f1 precedes the S phase
by approximately two hours. We interpret this to mean that
the phosphorylation of the old f1 histone begins before the
phosphorylation of the new. In fact, when the cells begin
to enter the S phase the phosphorylation rate becomes constant and then is accelerated again as cells approach G2.
This data will be shown tommorrow.

R. CHALKLEY: I have always shied away from that possibility, although. I know that it existed, because it is true
that what we are looking at in that last experiment I described is new histone. The reason I shied away from
it was because I guess what I was doing is applying
Occam's razor. I was optimistic that new histone would
be phosphorylated in a similar fashion, simply because

275

it would simplify the interpretation. We were in fact,
gearing up to try and look at all phosphorylation of
all histone at that time in the cell cycle in the manner
you described, because we recognized it was something that
had to be done. So far as our cell cycle data is concerned, we were using three hour pulses of ^{32}P, so I
do not think that we would be able to detect a shift of
two hours.

L.R. GURLEY: There is one large difference in our observations. In G2 cultures we observe an increase in fl phosphorylation while you find a decrease. In fact, we observe
fl phosphorylation in G2 to be double the rate in S phase.

R. CHALKLEY: I would not be surprised if phosphorylation
were to continue past S, since we are now thinking of it
in terms of a post-replicative effect, and that almost demands that our half hour lag must proceed beyond the end of
the S period. And I suppose you only need to argue that
different cells might have a different lag period, to be
able to see substantial measure of phosphorylation in G2,
so I do not regard that as a problem.

G.H. DIXON: Could I ask Dr. Gurley a question? This two
hour period of fl phosphorylation before the S-phase begins,
do you think this might be associated with unwinding the
chromotine in preparation for S-phase?

L.R. GURLEY: We have considered this possibility, but have
no evidence.

L.W. LONGTON: Aram Balejian and I have isolated a
histone from parotid fluid in the amount of about 1 mg per
100 ml of parotid solution. We have also found it in blood
serum at about 1 mg per 100 ml of histone. The histone in
the parotid was found associated with a glycoprotein. In
the blood, I am conjecturing now, perhaps it may have been
associated with the fibrinogen, but I am not sure about
that. We have taken collagen from cells and have found a
histone like material associated with it. We have indeed
isolated the histone, and determined its electrophretic
migration, we have found its amino acid composition, which
is consistant with histone, and yet it is in the fluid
compartments as differing from your intracellular histone.

R. CHALKLEY: Well, I guess all I can say is that it is a truly remarkable finding. If substantiated, it is going to make you a very famous man.

J. ROTH: Would you care to speculate on an apparent contradiction? Would you say that cyclic AMP is implicated in the phosphorylation or dephosphorylation of the histones?

R. CHALKLEY: I do not know.

J. ROTH: Since you noted an increase in the phosphorylation of the histones in tumor cells relative to nontransformed cells, and tumor cell proliferation has been shown to be inhibited by cyclic AMP, would a cyclic AMP dependent phosphatase be a reasonable control point in the system?

R. CHALKLEY: Well, you could always have a histone kinase that was turned off by phosphorylation.

T.A. LANGAN: I would like to comment on the previous comments. It seems to me that you have shown that the sites that are phosphorylated in your growing cells, are different from those phosphorylated by the cyclic AMP dependent activity, that one would expect these are cyclic AMP independent histone kinases that are involved in growth regulated phosphorylation. And also, as far as I am aware, maybe there are other people who know better, there are not fluctuations in the level of cyclic AMP through the cell cycle that would suggest that they are regulating the time of phosphorylation.

THE ROLE OF ENZYMIC MODIFICATION IN THE CONTROL OF HISTONE AND PROTAMINE BINDING TO DNA*

G.H. Dixon,   E.P.M. Candido and A.J. Louie

Extensive phosphorylation of protamine and both phosphorylation and ε-NH_2-acetylation of histones takes place in developing trout testis cells (1). Phosphorylation and the major part of the acetylation of histones occurs in stem cells and primary spermatocytes and is correlated with the synthesis of histones and DNA while protamine phosphorylation takes place in spermatid cells which are not synthesizing DNA (2). All the seryl residues in the protamines can be found in the phosphoryl form (3) while the sites of phosphorylation have been determined in histones IIb_1 and IV to be at the N-terminus in the identical sequence N-Ac-Ser(PO_4)-Gly-Arg (4). In trout testis IIb_2, a seryl residue at position 6 in the sequence - $(\varepsilon$-Ac)-Lys-Ser(PO_4) -Ala is phosphorylated (1). Histone III is also phosphorylated but the sites in trout testis have not yet been elucidated. However in calf thymus histone III, there appear to be two sites of phosphorylation at seryl residues 10 and 27 (5).

In sharp contrast to these N-terminally-located sites of

modification, testis very lysine rich histone I is phosphorylated predominantly near its C-terminus at Ser 157 in the sequence - Ala-Ala-Lys-Lys-Ser(PO_4)-Pro-Lys-Lys (4) although electrophoretic studies of whole histone I indicate that up to three other sites can be phosphorylated. Candido and Dixon (6, 7) have determined the sites of ε-amino acetylation in trout testis histones IIb_1, IIb_2, III and IV after labelling testis cells with ^{14}C-acetate; in every case the sites of acetylation are in the N-terminal quarter of the molecule as follows:

Histone IIb_1	Lys 5	(6)
Histone IIb_2	Lys 5, Lys 10, Lys 13, Lys 19	(7)
Histone III	Lys 9, Lys 14, Lys 18, Lys 23	(7)
Histone IV	Lys 5, Lys 8, Lys 12, Lys 16	(6)

Again histone I shows totally different behaviour in not being ε-NH_2-acetylated at all (6). Kinetic studies of acetylation of trout histone IV indicate that newly synthesized histone IV molecules are progressively acetylated at lysyl residues 5, 8, 12 and 16 and then deacetylated before being bound to DNA (8). Similarly, histone IIb_1 is phosphorylated at Ser 1 and acetylated at Lys 5 shortly after synthesis and probably before being bound to DNA.

It is suggested that these modifications in the most

basic regions of histone molecules may provide a mechanism whereby the very strong electrostatic binding of histones to DNA may be modulated so that the newly synthesized histone may find a kinetic route to its correct, final conformation in the tight complex with DNA.

1. G.H. Dixon, Karolinska Symposia on Research Methods in Reproductive Endocrinology, Gene Transcription in Reproductive Tissue (1972) p. 130.

2. A.J. Louie & G.H. Dixon, J. Biol. Chem. 247 (1972) 5498.

3. M.M. Sanders & G.H. Dixon, J. Biol. Chem. 247(1972) 851.

4. M.T. Sung & G.H. Dixon, Proc. Nat. Acad. Sci. (U.S.) 67 (1970) 1616.

5. W.F. Marzluff Jr. & K.S. McCarty, Biochemistry 11 (1972) 2672, 2677.

6. E.P.M. Candido & G.H. Dixon, J. Biol. Chem. 246 (1971) 3182 and 247 (1972) 3868.

7. E.P.M. Candido & G.H. Dixon, Proc. Nat. Acad. Sci.(U.S.) 69 (1972) 2015.

8. A. J. Louie & G.H. Dixon, Proc. Nat. Acad. Sci. (U.S.) 69 (1972) 1975.

*Supported by Medical Research Council and National Cancer Institute of Canada

DISCUSSION

B. COOPERMAN: In your model it seems that if the acetylation really were to be necessary, as you postulated it is, that there would be a substantial kinetic barrier for conformational changes between acetylated and non-acetylated forms so that it might be possible to see distinct conformational differences, between acetylated and non-acetylated forms in isolated histones. Do you see such changes?

G.H. DIXON: We have not done these experiments, we are waiting for some physical chemists to do them. We are really neither equipped nor competent to look at conformation of histones in solution. I would really very much like to see such data.

R.W. LONGTON: What is the fate of the histone after it is released by the protamine? Are there any evidences of a transport mechanism for extracellular delivery?

G.H. DIXON: There is really no hard data on it, but I think it depends considerably on the species. There seems to be some species where one can see extrusion of granules of material that look as if they might be basic protein, but nobody has shown chemically that they are histones. The other possibility that we looked into in length in some work by Dr. Marilyn Sanders, was to see if there was proteolytic degradation of histones. In some work of Dr. Marushige, in the nucleoprotein at the stage of replacement, we could see a number of what looked like histone fragments. Dr. Sanders was able to detach a proteolytic activity from the chromatin which we called histonase and although her work is not complete, it would seem that this proteolytic activity is able in fact to lop off the N-terminal 20 amino acids of two and probably three of the histones. Thus there does seem some possibility of proteolysis after removal of the histones.

R.W. LONGTON: Aram Balekjian and I have discovered and isolated a fluid or extracellular histone, both from the parotid fluid and from the blood serum. It is present in about 1 mg per 100 ml fluid and this work is in press presently. There appears to be three chromatographic peaks of histone fraction from the parotid fluid and there appears to be two peaks of histone fractions from the serum. In the serum there is a phosphorylated peak and a de-phosphorylated or non-phosphorylated peak. Again this is new work that we have been involved with and we are looking for possible functions for this.

G.H. DIXON: Do I understand you in regard to these histones that you are suggesting that they have come off the nuclear histone and have been secreted into the extracellular fluid compartment?

R.W. LONGTON: Well, we do not know if they are secreted. We find the association of the histones with the glycoprotein of the parotid. In the serum of course, we have not found this association. However, when we looked at collagen, there was a histone-like material that perhaps is associated with collagen, so perhaps, a glycoprotein may be involved in the transport mechanism, but this is very preliminary.

V.G. ALLFREY: I just wanted to comment briefly on the question raised by Dr. Cooperman concerning whether histone acetylation makes a difference in the conformation of a histone. This has been answered directly by separating histone F2A1 subfractions which differ in their degree of acetylation. The separation is based on differences in the chromatographic properties of the acetylated and non-acetylated forms of F2A1 on carboxymethyl-cellulose. Drs. Lawrence Wangh and Adolfo Ruiz-Carrillo have employed CMC-column chromatography to prepare F2A1 subfractions containing 0, 1, or more epsilon-N-acetyllysyl residues. These were then compared for their effects on DNA-binding, using circular dichroism as a measure of conformational change. The CD studies were carried out by Drs. Alice Adler and Gerald D. Fasman of Brandeis University. They observed enormous differences, as far as the conformation of the DNA-histone complex was concerned, depending upon whether the histone did or did not contain acetyl groups.

K.S. McCARTY: Do you have any evidence, (using for example reporter-dye or actinomycin D) to establish the location of protamines within the major or minor groove of the DNA?

G.H. DIXON: No, we have not done these experiments. The

literature says that they are in the minor groove, this conclusion being based on some X-ray crystalography, done in Wilkin's group a number of years ago. I do not know that anyone has reexamined this, it would be very well worth doing I think.

K.S. McCARTY: Would it be useful to construct a helical wheel for protamines as for histones?

G.H. DIXON: Yes, it would. We must keep in mind the fact that there is tremendous condensation of chromatin. This almost implies that protamine has to cross-link from one region of DNA to another, so I doubt that it can lie completely along either groove. It may do for a short region and then it might cross-link to another major groove. This is all speculative and needs to be reexamined.

E.G. KREBS: Do you want to comment on the role of small molecules, effectors such as cyclic AMP, calcium or other components which may play a part in regulating the various reactions you described?

G.H. DIXON: The chain of regulation is sketchy. It is known from work of Menon and Smith in Vancouver that if one treats a testes cell suspension with gonadotropin, there is undoubtedly an increase in cyclic AMP. When I last discussed the work, they did not know in which cells this was occurring but one of the theories of gonadotropin action says that there may be an effect on the replication of stem cells. The gonadotropin is thought to have a mitogenic effect in generating more and more stem cells and that one might be accumulating a number of hits by gonadotropin which perhaps has the effect of steadily increasing cyclic AMP level to the point where one induces cell division and replication. Unfortunately, the data with cells in culture indicates that low cyclic AMP levels are correlated with increased cell division so there is a mystery here. Of course at the other end of the spectrum we have seen that there is a protamine kinase which is, as you know, cyclic AMP dependent and I showed that in vivo protamine is certainly phosphorylated. We do not know how the acetylation is regulated. One of the very early experiments we did of course was to see if the chromatin-bound histone acetylase was sensitive to cyclic AMP. It

seemed not to be, and we were very disappointed. I think
this needs to be looked at again since there are a number
of possibilities. For example, effectors which perhaps
are in turn controlled by gonadotropin.

HISTONE PHOSPHORYLATION AND REGULATION OF

NUCLEAR FUNCTION

Thomas A. Langan

Non-dividing, non-growing cells, for example the cells
of adult rat liver, contain abundant protein kinase activity capable of catalyzing the phosphorylation of lysine-
rich and other histones (1). These enzymes are found
mainly as soluble proteins distributed throughout the cytoplasm and the nucleoplasm; in addition substantial activity
is found bound to extensively washed and purified chromatin.

In contrast to the massive phosphorylation of histones
which occurs during S phase in rapidly growing cells (2)(3)
and during spermatogenesis in trout testis (4), the phosphorylation of lysine-rich fl histone in adult rat liver
involves a very small fraction of the total fl histone.
In particular, the phosphorylation of rat liver fl histone
which is regulated by cyclic AMP and hormones (5)(6) appears limited to about one percent of the total fl histone.
This phosphorylation occurs at a specific site in the histone molecule, which in at least one case is known to be

distinct from the site involved in non-cyclic AMP-dependent phosphorylation (1).

The differences in histone phosphorylation occurring in dividing and non-dividing cells suggest that the nature and extent of histone phosphorylation may be related to the nature and extent of the genome manipulation required in a given cellular situation. In dividing cells and during spermatogenesis the entire genome is manipulated in a non-gene specific manner to provide for replication or chromosomal protein replacement. Under these conditions, histone phosphorylation is massive, involving histone associated with the entire genome. On the other hand, the very limited extent of phosphorylation we have observed in our studies of cyclic AMP regulated histone phosphorylation in non-dividing cells is more in keeping with the selective manipulation of the genome which would be required for the activation of specific gene transcription. The fact that distinct sites in f1 histone have been found to be involved in cyclic AMP-dependent and independent phosphorylation further suggests that different types of histone phosphorylation may serve different functions in the cell.

We have investigated the effect of phosphorylation of

specific sites in lysine-rich histone on the interaction
of the histone with DNA and on the restriction of template
activity of chromatin. In collaboration with Fasman and
co-workers (7)(8), phosphorylation of lysine-rich histone
at either of two specific serine residues has been found
to markedly reduce the ability of the histone to cause
structural changes in double stranded DNA, while phosphorylation of both serine residues causes much larger changes.
Doubly phosphorylated lysine-rich histone is also less effective than non-phosphorylated histone in restricting RNA
synthesis on reconstituted chromatin templates. These observations provide suggestive although not conclusive evidence that, in non-dividing cells, histone phosphorylation
may have a function in the regulation of gene activity.

(1) T. A. Langan, Ann. N. Y. Acad. Sci. 185 (1971) 166.

(2) R. Balhorn, W. O. Rieke and R. Chalkley, Biochemistry
    10 (1971) 3952.

(3) R. Balhorn, J. Bordwell, L. Sellers, D. Granner and
    R. Chalkley, Biochem. Biophys. Res. Comm. 46 (1972)
    1326.

(4) A. J. Louie and G. H. Dixon, J. Biol. Chem. 247 (1972)

(5) T. A. Langan, J. Biol. Chem. 244 (1969) 5763.

(6) T. A. Langan, Proc. Natl, Acad. Sci. 64 (1969) 1276.

(7) A. J. Adler, B. Schaffhausen, T. A. Langan, and G. D.
    Fasman, Biochemistry 10 (1971) 909.

(8) A. J. Adler, T. A. Langan and G. D. Fasman, Arch.
    Biochem. Biophys. In press.

DISCUSSION

G.H. DIXON: In some experiments that Jergil Sung and I did
a couple of years ago in which we collaborated with you in
exchanging kinases, it seemed that the source of the kinases
(we were comparing the trout testis kinase with you liver
kinase) seemed to be relatively unimportant. But it seemed
that the site which you saw phosphorylated in the liver,
namely a serine 37 or 38, depending on your numbering, was
still phosphorylated in the liver histone by either kinase
and the site which we saw in trout histone I <u>in vivo</u> which
is in the C-terminal region serine 157, was still phosphorylated in the trout histone <u>in vitro</u> are by either kinase.
This seems to imply that there was some kind of different
presentation of the surface of the histone to the enzymes
in two cases. Perhaps, this was due to some actual difference between the conformation of the two histones. Do you
have any comments on that? Have you thought anymore about
that experiment?

T.A. LANGAN: I have thought a lot about it, but I really
do not have any comments that I think would get us any
further than we are.

G.H. DIXON: It is odd because the sequence of trout histone
f_1 which has been done in my laboratory by Mr. A.S. MacLeod,
seems not to be really very different from David Cole's
rabbit mammary gland f_1.

T.A. LANGAN: Did you make the suggestion yourself that the
configuration of the histone could have quite an effect on
the phosphorylation that occurred?

G.H. DIXON: Yes, we made that suggestion, but the emerging
sequence does not seem to give us much basis for markedly
different conformations.

T.A. LANGAN: A change in a particular place might have a
surprising and unexpected large effect.

G.H. DIXON: Yes, I agree.

G. VIDALI: Does the kinase work on the fragment of the
molecule?

T.A. LANGAN: Yes, we have studied the action of the kinases on histone split at the tyrosine residue that was in that scheme and we have also split histone that was already phosphorylated and that when you do that you find the pattern of phosphorylation that agrees with the location of these sites and in addition the separate pieces are fairly good substrates as well.

C. MOORE: Since both Dr. Langan and Dr. Dixon are up on the stage, I am wondering if there is any difference in ideology with regards to the mode of interaction between histones and DNA. Earlier today we were fortunate enough to hear Dr. Dixon talk about the acylation of the lysine groups and the meaning or the significance of the acylation and deacylation as far as interaction between the histones and DNA is concerned. We just heard Dr. Langan talk about the requirement for phosphorylation to facilitate this same interaction without any mention or overt concern about the presence or absence of acylated lysine groups. Are you two gentlemen in agreement. Are you both correct or are your ideas mutually exclusive?

T.A. LANGAN: I would answer by saying that it seems perfectly all right to me that you can effect the interaction of the histone with DNA by either a acylation or phosphorylation to begin with and that these things might happen under different conditions and have different consequences.

G.H. DIXON: If you look into the detailed kinetics of phosphorylation in our system, the phosphorylation of each histone seems to show very definitely different kinetics. In other words you can say quite clearly that histone IIb (F2 a2) is rapidly phosphorylated, and this is at the N-terminal site, whereas with histone IV and histone 1, there is a considerable lag period of several hours before they get phosphorylated. The kinetic curves of phosphorylation and dephosphorylation of these histones are really quite distinct and so this brings up a really interesting question of whether there are specific kinases for each of these histones since the kinetics are so different, or whether this represents some sort of different exposure to a common kinase as Dr. Langan has suggested, due to changes in the coiling or uncoiling or supercoiling of the chromosome. I think this is an area that really needs to be looked into

since I do not think you can generalize for all of the
histones at any one time for either phosphorylation or
acetylation. They each seem to be a separate case. They
each have their own kinetics.

T.A. LANGAN: It seems to me that if you look at the different types of histone modification that go on, and in
particular if you compare phosphorylation under different
situation, such as spermatogenosis or growing cells as we
will certainly hear in a few minutes, or in non-dividing
cells under hormonal control. You can see that the extent
of histone modification is more or less in keeping with the
fraction of the genome that you know. That is, if a cell
has to replicate its chromatin, this is a non-gene specific
manipulation of the entire genome and so you have massive
histone modification under those conditions. Similarly,
when you replace the entire protein complement associated
with the DNA and spermatogenesis, you have massive modifications, but here in the non-dividing cell the extent of
phosphorylation is much less. It is apparently selective
and it suggests that in each case you are selectively
manipulating a small fraction of the genome. That is why
one sees a lesser amount of phosphorylation. This whole
idea also implies that if you modify histone function by
chemical modification of the histone, when you make these
different modifications at different times, you are modifying different functions of histones. It suggests that
histones have perhaps multiple functions in the course of
not only gene activation, but structural roles in replication as well.

FREE COMMUNICATIONS

IDENTIFICATION OF THE F1-HISTONE PHOSPHOKINASE OF MITOTIC CHINESE HAMSTER CELLS

Robert S. Lake, Lab. Biol. Viruses, Nat. Inst. of Allergy and Inf. Diseases, NIH, Bethesda, Md.

Metaphase-arrested (M) Chinese hamster (CH) cells exhibit a higher level of F1 (lysine-rich) histone phosphorylation(1) and F1-specific phosphokinase activity (2) than interphase (I) cells. Two major phosphokinase activities, designated KI and KII, were extracted from whole exponentially growing cells with 0.35M NaCl and separated in 0.2M NaCl by Sephadex G-200 gel filtration. KI ($\sim$250,000 da) is cyclic 3',5'-adenosine monophosphate (cAMP)-dependent at pH 7.0 and uses F1-histone as substrate at pH9.0 but prefers phosvitin at lower pH values. KII ($\sim$90,000 da) is cAMP-independent and prefers F1-histone to all other substrates at pH between 6.5 and 9.0. KII also differs from KI in its K_m for ATP, in vivo decay in the presence of cycloheximide, and pattern of in vitro derived F1-phosphopeptides.

Comparative examination of M and I cells for these two activities reveals that KII is responsible for the overall high activity in M cells. Pulse labelling of CH cells with ^{32}P during traverse of the G2-M boundary reveals a F1-tryptic phosphopeptide pattern unique to M cells. Seven major peptides derived by in vitro phosphorylation of whole unfractionated F1 with the KII enzyme correspond to these M cell specific phosphorylation sites observed in vivo. It is concluded that both KI and KII can act on F1 throughout interphase, but that KII predominates during mitosis.

REFERENCES

(1) Lake, R. S., Goidl, J. A., and Salzman, N. P.
    Exp. Cell Res., 73, 113 (1972)
(2) Lake, R. S. and Salzman, N. P. Biochemistry
    In the press (1972)

THE INDEPENDENCE OF HISTONE PHOSPHORYLATION FROM DNA SYNTHESIS

Lawrence R. Gurley, Ronald A. Walters, and Robert A. Tobey,
Biomedical Research Group, Los Alamos Scientific Laboratory, University of California, Los Alamos, New Mexico   87544

There has been considerable interest in the hypothesis that a modification of histones by phosphorylation may weaken the interaction between histone and DNA, resulting in an activation of the DNA template. However, correlations were found recently between DNA synthesis and histone phosphorylation which suggested that histone phosphorylation occurs only at a time when DNA is synthesized. These experiments raised questions concerning the involvement of histone phosphorylation in control of gene activity. Since these conclusions were different from those we had reached previously in x-irradiation experiments, we have investigated the phosphorylation of electrophoretically pure histones in synchronized Chinese hamster cells. We find that phosphorylation of histone f2a2 is independent of cell-cycle position, occurring in the G_1, G_2 and M phases when DNA synthesis is absent, as well as in the S phase when DNA synthesis is active. In contrast, phosphorylation of histone f1 was found to be absent in G_1 cells. Histone f1 phosphorylation was found to be active during the S phase, and the rate of f1 phosphorylation increased over the S-phase rate in G_2 and M. These results indicate that phosphorylation of histone f2a2 is independent of f2a2 synthesis, independent of DNA synthesis, and independent of f1 phosphorylation. Because f2a2 is actively phosphorylated in G_1 cells known to be active in RNA synthesis as well as in S, G_2 and M cells, we feel that phosphorylation of f2a2 should continue to be considered in models concerning activation of DNA template activity. The active phosphorylation of f1 during the G_2 and M phases, as well as during the S phase, indicates that f1 phosphorylation is not strictly an S-phase phenomenon involved with DNA synthesis. Rather, it suggests that f1 phosphorylation is a more generalized late-interphase event which begins at the S phase, possibly in preparation for cell division. (This work is being performed under the auspices of the U. S. Atomic Energy Commission)

IN VITRO PHOSPHORYLATION BY ATP AND GTP OF VINBLASTINE-ISOLATED MICROTUBULES FROM CHICK EMBRYO MUSCLES

M.M. Piras and R. Piras, Instituto de Investigaciones Bio-químicas-Fundación Campomar, Buenos Aires (28), Argentina and Instituto Nacional de Farmacología y Bromatología.

Protein phosphokinase (PrK) decreases abruptly in embryonic chick muscle during the 11 to 17th. day period (1), concomitantly with the decrease of the mitotic index of this differentiating tissue. Various PrK can be detected, some of them being typical of the embryonic or the adult muscle (2). Microtubules, which are both cuantitatively and functionally prominent in the embryonic tissue, have now been found to be one of the natural substrates for an embryonic PrK.

The microtubular fraction (MF), isolated with 1 mM Vinblastine from 11-days old embryos, binds ^{3}H-colchicine, has PrK activity, and more than 90% of its total protein migrates as two closely associated bands of MW 53000 in SDS-polyacrylamide gel electrophoresis. Incubation of the MF with $(\gamma-^{32}P)$-ATP and Mg^{2+} results in a rapid incorporation into a 5% TCA precipitable fraction, which is RNAse and DNAse resistant and pronase sensitive. ^{32}P-Ser is liberated upon acid hydrolysis. Most of the incorporated radioactivity migrates with the faster moving band on SDS-gel electrophoresis and emerges on Biogel P-300 coinciding with the colchicine binding (MW 120 000), but separated from the PrK activity (MW34 000). Phosphorylation by ATP (K_m = 8 μM) requires Mg^{2+} or Mn^{2+}, and is inhibited by salts and Vinblastine. GTP can replace ATP in the phosphorylation of the MF.

The results described, as well as the in vivo incorporation of ^{32}P obtained into an embryonic MF and the lack of in vitro incorporation into an adult muscle MF, suggest that microtubule phosphorylation might be a mechanism of controlling the function and/or assembly of this important cellular structure during muscle differentiation.

REFERENCES

(1) M.M. Piras, R. Staneloni, B. Leiderman and R. Piras, FEBS Letters 23 (1972) 199.
(2) R. Piras, R. Staneloni and M.M. Piras, Abstracts of VIIIth Meeting Soc. Arg. Invest. Bioq. (1972) 23.

This research was supported in part by grants from the Consejo Nacional de Investigaciones Científicas y Técnicas and the Instituto Nacional de Farmacología y Bromatología.

ADENOSINE 3':5'-CYCLIC MONOPHOSPHATE MEDIATED CORTISOL
INDUCTION OF HeLa ALKALINE PHOSPHATASE. M.J. Griffin, G.H.
Price and S. Tu , Oklahoma Medical Research Foundation,
Oklahoma City, Oklahoma.

HeLa 65 cells cultured with 3 μM cortisol possess a more
catalytically efficient alkaline phosphomonoesterase than
do controls, although the number of enzyme molecules per
cell is similar for the two cell types. This membrane glycoprotein enzyme has been solubilized by n-butanol extraction
and purified to homogeneity by preparative disc-gel electrophoresis or isoelectric focusing. In seven independent
experiments, the control enzyme was found to have 12 ± 1
moles inorganic phosphate covalently bound per mole protein
compared to 4 ± 0.3 for the cortisol-induced species.
Dibutyryl cyclic AMP (1 mM) elicits partial enzyme induction
in dividing cells and the purified enzyme has 8 moles of
phosphate per mole enzyme, a value intermediate between the
phosphate content of control and cortisol induced enzymes.
Hence, increased activity effected by either cortisol or
the cyclic nucleotide analogue is associated with a decrease
in enzyme phosphate content.
    Intracellular concentrations of cyclic AMP (cAMP) have
been measured by a protein binding assay. Whereas HeLa 65
control has 0.39 ± 0.17 pmoles/10^6 cells, cells cultured
for 24 hrs with steroid have 14 ± 3 pmoles cAMP/10^6 cells
and for 48 hrs or 72 hrs, 2.2 ± 0.3 pmoles cAMP/10^6 cells.
The enzyme induction begins after 24 hrs and plateaus at
70 hrs. Continuous subculture with glucocorticoid in the
growth medium (Hcr cells) results in continued induced
activity and 1.1 ± 0.2 pmoles cAMP/10^6 cells. HeLa 71 has
constitutive enzyme activity which is 30 fold increased
relative to enzyme from HeLa 65 Hcr. This strain has 4.2 ± 0.4 pmole cAMP/10^6 control cells. These data are consistent with a mechanism in which cortisol stimulates adenyl
cyclase and/or inhibits cAMP phosphodiesterase of HeLa 65,
resulting in increased cellular concentrations of cAMP
which effects dephosphorylation (or inhibits phosphorylation)
of alkaline phosphatase, a process that converts a less
active enzyme to a more active species.
    Recently, HTC cells have been reported to have a three-
fold increase in cAMP content when cultured with 1 μM
dexamethasone [V. Manganiello and M. Vaughan, J. Clin. Invest.
51 (1972) 2763].

CYCLIC AMP ACTIVATED PROTEIN KINASE OF KIDNEY MEMBRANES

Leonard R. Forte and Keith H. Byington, The Department of Pharmacology, School of Medicine, University of Missouri, Columbia, Missouri 65201

Kidney plasma membranes prepared by the method of Fitzpatrick et al. (1) exhibit hormone and NaF activated adenylate cyclase and phosphodiesterase activities (2). The receptor system for intracellular cAMP has been proposed to be a soluble cytoplasmic (cytosol) protein kinase (3). The present study compares the properties of a plasma membrane and cytosol protein kinase prepared from hog kidney cortex. Examination of the kinetics of binding of ^{3}H-cAMP to these preparations indicated that cytosol contained one binding site which was saturated at 5×10^{-7}M ^{3}H-cAMP whereas the plasma membranes exhibited characteristics of multiple binding sites which were not saturated with concentrations as high as 1×10^{-5}M. Competitive binding studies showed that both preparations had high affinity for cAMP, N6-monobutyryl cAMP and 8-bromo-cAMP and relatively low affinity for N6, O2-dibutyryl cAMP, O2-monobutyryl cAMP, cGMP and 2'3' cAMP. Binding of ^{3}H-cAMP to plasma membranes was abolished or diminished by heating, treatment with trypsin, protease and neuraminidase but not affected by treatment with phospholipase-C, detergents or NEM. Both the cytosol and plasma membrane preparations exhibited cAMP activated protein kinase activity. Activation was shown over the cAMP concentration range of 5×10^{-9} to 1×10^{-6}M with a 2-2.5 fold increase in protein kinase activity at maximal activation levels of cAMP. A general correlation was found between the relative binding affinities of the above nucleotides and the concentration of nucleotide required for activation of protein kinase of plasma membrane and cytosol preparations. This study shows that renal cyclic AMP dependent protein kinase is associated with both plasma membrane and cytoplasmic fractions. The membrane protein kinase system may be involved in hormonal regulation of membrane transport.

REFERENCES
1. D.F. Fitzpatrick, G.R. Davenport, L.R. Forte and E.J. Landon, J. Biol. Chem. 224:3561 (1969).
2. L.R. Forte, Biochim. Biophys. Acta 266:524 (1972).
3. D.A. Walsh, J.P. Perkins and E.G. Krebs, J. Biol. Chem. 243:3763 (1968).

This research was supported by grant AM-14787 from the National Institutes of Health, U.S. Public Health Service.

CYCLIC NUCLEOSIDE PHOSPHATE DEPENDENT PROTEIN KINASE IN
SEA URCHIN GAMETES AND EMBRYOS

M.Y.W. Lee and R.M. Iverson,Laboratory for Quantitative
Biology, University of Miami, Miami, Florida.

A protein kinase which phosphorylates histones and
protamine was shown to be present in the sperm, oocyte
and developing embryo of the sea urchin, Lytechinus
variegatus.   Levels of protein kinase in the sperm (units
per mg soluble protein) were 30-fold higher than those in
the oocytes or gastrula embryos, and were at least 100-fold
higher than the levels reported in various tissues and
species of Arthropoda by Kuo et al. (J. Biol. Chem. 246,
7159, 1971).   The egg and sperm protein kinase activities
have similar pH optima and a similar dependence on either
cyclic AMP or cyclic GMP.   In view of the widening
implication of protein kinases in the regulation of cellular processes, the relatively high levels of this enzyme
in the sea urchin sperm are suggestive that it may have a
role in gametogenesis or the fertilisation process.   We
have therefore characterised the sea urchin sperm protein
kinase.   The enzyme was purified by calcium phosphate gel
extraction, DEAE cellulose and Sephadex G-200 chromatography to a specific activity of 390 nmoles phosphate
transferred from ATP to protamine/min./mg protein.   The
enzyme has a molecular weight of about 290,000 by Sephadex
chromatography.   The cyclic AMP and cyclic GMP dependent
activities appeared to be associated with the same protein
species. Mg^{++} was essential for enzyme activity and the
rate of protamine phosphorylation was stimulated 2.5-fold
by an optimal concentration of 0.9M NaCl.   The K_m for
ATP (minus cyclic AMP) was 0.119 mM $\pm$ 0.013 (S.D.) and
0.055 mM $\pm$ 0.009 (S.D.) in the presence of cyclic AMP.
The specificity of the protein kinase toward a series of
protein substrates was also examined.   Preliminary studies
of protein kinase levels during the early development of
sea urchin embryos have revealed a consistent temporal
fluctuation of enzyme levels which appears to be associated
with the cell cycle.

[Supported by NIH training grant HD00187 (M.Y.W.L.) and
NSF grant GB 7709]

PROTEIN PHOSPHORYLATION IN RETINAL PHOTORECEPTORS

Richard G. Pannbacker, C. F. Kettering Research
Laboratory, Yellow Springs, Ohio

Two distinct protein kinase activities are
found in mammalian photoreceptors. These enzymes
would appear to respond to light in opposite ways.
Rhodopsin kinase activity in intact rod
outer segments has been shown to be light-dependent (1,2). When bovine rod outer segments were
lysed by French pressure cell treatment and separated into soluble and membrane fractions, we
found that rhodopsin kinase activity was membrane
bound. Phosphorylation of rhodopsin in this
preparation was light-dependent and was not
affected by cyclic nucleotides. Rhodopsin phosphorylation in incubated rabbit retinas was also
found to be light-dependent.
Cyclic nucleotide-dependent protein kinase
was found in the soluble fraction prepared from
bovine rod outer segments. With histone as the
substrate, the reaction was half-maximally stimulated by $5 \times 10^{-8}M$ cyclic AMP. Cyclic GMP was a
less effective stimulant. This enzyme did not
phosphorylate rhodopsin, but did phosphorylate a
soluble protein. The observed inhibition of rod
outer segment adenyl cyclase and guanyl cyclase
by light (3,4) suggests that the activity of the
soluble protein kinase would be decreased on
bleaching in the intact cell.

REFERENCES

(1) D. Bownds, J. Dawes, J. Miller and M.
    Stahlman, Nature New Biol. 237 (1972) 125.
(2) H. Kuhn and W. J. Dreyer, FEBS Letters 20
    (1972) 1.
(3) M. W. Bitensky, R. E. Gorman and W. H. Miller,
    Proc. Nat. Acad. Sci. U.S. 68 (1971) 561.
(4) R. G. Pannbacker and D. E. Fleischman,
    Biophys. Soc. Abstr. (1972) 207a.

This investigation was supported by NIH research
grant #EY00860 from the National Eye Institute.

cAMP DEPENDENT PROTEIN KINASE-CATALYZED PHOSPHORYLATION OF CARDIAC MICROSOMES

M.A. Kirchberger, M. Tada, S. Yoshioka, and A.M. Katz. Departments of Medicine and Physiology, Mount Sinai School of Medicine of the City University of New York, N.Y., N.Y.

Ca-uptake by dog cardiac microsomes is more than doubled by partially purified bovine cAMP-dependent protein kinase (1). These microsomes, an enriched sarcoplasmic reticulum (SR) fraction, also contain epinephrine-sensitive adenylate cyclase (ACase). To characterize further cAMP-dependent protein kinase mediation of Ca movements, protein kinase (PK)-catalyzed phosphorylation of cardiac microsomes, made by a slightly modified method of Harigaya and Schwartz (2), was studied.

Microsomes (0.5 mg/ml) were incubated at 30 C in the medium used to measure Ca-uptake: 40 mM histidine buffer, pH 6.8, 0.12 M KCl, 2.5 mM Tris-oxalate, Ca-EGTA buffer ($[Ca^{++}]$=0.75 μM), cAMP and/or PK. Reactions were started with 5 mM MgATP[γ-^{32}P] and stopped with 10% TCA + 0.1 mM KH_2PO_4. Phosphorylated microsomes were washed four times with the same solution and counted.

Microsomal phosphorylation was approximately tripled in the presence of 0.5 μM cAMP + PK. In either cAMP or PK alone, phosphorylation appeared to depend on endogenous PK or ACase activity, respectively. In microsomes with high PK activity, endogenous cAMP production stimulated as much, but less rapid, phosphorylation as did 0.5 μM cAMP + 0.1 mg/ml partially purified PK. Epinephrine (0.5 mM) which can be shown to increase microsomal ACase, increased phosphorylation rate 1 1/2 to 2-fold in the presence of added PK. Slower enhancement of phosphorylation by epinephrine seen in the absence of added PK was due presumably to endogenous microsomal PK.

These findings suggest that ACase and PK activities of cardiac SR may participate in the regulation of intramyocardial Ca movements. Such a regulatory system, localized at the SR, may mediate the enhanced contractility induced by epinephrine.

REFERENCES
(1) M.A. Kirchberger, M. Tada, D.I. Repke, and A.M. Katz, J. Mol. Cell. Cardiol. 4 (1972) in press.
(2) S. Harigaya and A. Schwartz, Circ. Res. 25 (1969) 781.

SUBSTRATE-SPECIFIC PHOSPHORYLATION OF RIBOSOMAL PROTEINS FROM RABBIT RETICULOCYTES

J. A. Traugh, M. Mumby, and R. R. Traut, Department of Biological Chemistry, University of California, School of Medicine, Davis, California.

Three peaks of protein kinase activity have been resolved by $(NH_4)_2SO_4$ fractionation and DEAE-cellulose chromatography of the ribosome-free supernatant from rabbit reticulocytes. Peaks I and II, eluting at 0 and 0.1 M KCl respectively, are characterized by the preferential phosphorylation of basic substrates. Peak III, eluting at 0.2 M KCl, preferentially phosphorylates acidic substrates, and unlike Peaks I and II, can use $(\gamma-{}^{32}P)GTP$ as well as $(\gamma-{}^{32}P)ATP$ as a phosphate donor.

Specific ribosomal proteins in the 40S and 60S subunits of rabbit reticulocytes are phosphorylated by each of the protein kinase fractions. One protein band in the 40S subunit and six protein bands in the 60S subunits are phosphorylated by Peaks I and II as demonstrated by one-dimensional polyacrylamide gel electrophoresis in sodium dodecyl sulfate. A single protein band in the 40S subunit and four protein bands in the 60S subunit are phosphorylated by Peak III activity with either ATP or GTP. Only one of the 60S bands is coincident with those phosphorylated by Peaks I and II.

ACKNOWLEDGEMENTS

Supported by grants from the American Heart Association (69 461) and Damon Runyan Memorial Fund (DRG-1140).

A NOVEL, CYCLIC NUCLEOTIDE-DEPENDENT, PROTEIN KINASE FROM HUMAN POLYMORPHONUCLEAR LEUCOCYTES

P.-K. Tsung, N. Hermina and G. Weissmann, Department of Medicine, New York University School of Medicine, New York, N. Y.

To determine how cyclic nucleotides exert their effects on phagocytic cells (1), we have studied a protein kinase from human polymorphonuclear leucocytes (PMNL). The enzyme, after $(NH_4)_2SO_4$ precipitation and DEAE-cellulose purification had a pH optimum of 6.5, required Mg^{++}, Co^{++}, or Mn^{++}, and preferentially phosphorylated lysine-rich histone. It was activated equally by cIMP and cAMP (apparent Km's=62,88 nM), but far less by other cyclic nucleotides. Unlike cGMP-dependent kinases in other cells, the activities dependent upon cIMP and cAMP would not be separated by DEAE-cellulose, Sephadex G-75 or 200 chromatography, isoelectric focusing, or sucrose gradients. Neither deaminase nor transaminase activity accounted for the lack of cyclic nucleotide discrimination. Sephadex G-75 and gel electrophoresis showed $>$ 80% of kinase activity present as two proteins (M.W.: $\cong$ 60,000; 45,000) which could be dissociated still further to one band (M.W. $\cong$ 30,000) in the presence of $5\mu M$ cIMP. The smallest protein remained cyclic nucleotide-dependent. Studies of ^3H-cIMP or ^3H-cAMP binding showed that preparations from polymorphs were only 1/20 as active as from beef heart (at equal kinase activity). Moreover, the lysosomal fractions of PMNL (but not other blood cells) contained a heat-labile inhibitor of cyclic nucleotide binding by beef heart protein kinase preparations. The data suggest that the protein kinase (active on histone) of PMNL exists in multiple forms, is cyclic nucleotide-dependent in the dissociated form, and that since PMNL lysosomes influence cyclic nucleotide binding, the protein kinases of PMNL may not require the usual R subunit.

Reference

(1) Weissmann, G., Dukor, P. and Zurier, R.B., Nature New Biol., 231:131, 1971.

HISTONE PHOSPHORYLATION IN NORMAL AND REGENERATING RAT LIVER

R.E. Branson, J.L. Irvin, Department of Biochemistry, University of North Carolina School of Medicine, Chapel Hill, North Caroline

Studies on histone phosphorylation in rat liver indicate that 22 hours after partial hepatectomy ^{32}P is incorporated into histone fractions F_1, F_{2b}, F_{2a1}, and possibly F_3 and F_{2a2}. In normal rat liver, however, there appears to be a constant level of phosphorylation as measured by ^{32}P incorporation into fraction F_{2a2}, while the remaining histones incorporate little or no ^{32}P. Fractionation of F_1 by column chromotography (1) shows that three main fractions of F_1 incorporate ^{32}P during liver regeneration, while very little is incorporated into F_1 of normal liver.

Data will be reported on the effect of pharmacologic doses of hydrocortisone upon histone synthesis and phosphorylation in normal and regenerating liver.

REFERENCES

(1) J.M. Kinkade and R.D. Cole, J. Biol. Chem. 241 (1966) 5790.

CYCLIC AMP AND GROWTH CONTROL IN BHK CELLS TRANSFORMED BY THE POLYOMA VIRUS MUTANT TS-3

Wilfried E. Seifert, Salk Institute, San Diego, California

In ts-3-BHK cells certain properties (agglutinability, lack of topoinhibition) that are characteristic for transformed cells, are rendered temperature-dependent (1, 2).

Growth curves at different serum concentrations (without medium change and with daily medium change) at 31° and 39° demonstrate a higher serum requirement of ts-3-BHK at 39°.

The extent of growth inhibition by DBcAMP and theophylline depends both on the concentrations of these drugs and of the serum. Conditions were found where ts-3-BHK with respect to this inhibition behave like normal BHK at 39°; however, at 31° like the wild-type transformed Py-BHK.

Thus ts-3-BHK cells reflect the same temperature dependence in their serum requirement and in their sensitivity towards DBcAMP as was shown for their agglutinability (2). All these ts properties may well be the consequence of a single surface change caused by the ts-3 gene.

The cellular levels of cAMP (determined by the Gillman binding assay at different times of the growth curves in 2% and 10% serum at 31° and 39°) depend on the growth state of the ts-3-BHK cells. Only at later times when they had stopped growing at 39° (less than 5% labeled nuclei in autoradiography) cAMP was higher (factor 1, 5-2) at 39° than at 31°, while Py-BHK showed no significant differences.

REFERENCES

(1) R. Dulbecco, W. Eckhart, Proc. US Nat. Acad. Sci. 67, 1775 (1970).

(2) W. Eckhart, R. Dulbecco, M. M. Burger, Proc. US Nat. Acad. Sci. 68, 283 (1971).

This work was carried out in Dr. R. Dulbecco's laboratory and supported by NIH Grant No. CA-07592.

IDENTIFICATION OF THE CYCLIC AMP BINDING PROTEIN IN HUMAN ERYTHROCYTE GHOSTS BY COVALENT INCORPORATION OF A DIAZOMALONYL DERIVATIVE OF CYCLIC [3H] AMP.

C. E. Guthrow, Jr., D. J. Brunswick, B. S. Cooperman and H. Rasmussen, Departments of Biochemistry and Chemistry, University of Pennsylvania, Philadelphia

It has been shown that human erythrocyte ghosts contain a cyclic AMP binding protein and an endogenous, membrane-associated cyclic AMP-dependent protein kinase which catalyzes the phosphorylation of an approximately 40,000 molecular weight endogenous membrane protein(1). It has also been shown that a diazomalonyl derivative of cyclic AMP is specifically incorporated upon photolysis into the cyclic AMP binding site of rabbit muscle phosphofructokinase, and it was suggested that such derivatives could be used to isolate and identify cyclic AMP receptor sites (2).

We now report that the N^6-(Ethyl 2-diazomalonyl) derivative of cyclic [3H] AMP is specifically incorporated upon photolysis at 253.7 nm into human erythrocyte ghosts. It has been possible to use this label to identify the cyclic AMP receptor site in this system. SDS disc gel electrophoresis of solubilized ghost showed that one protein is labeled. Furthermore, the labeled protein has a molecular weight of approximately 40,000 and is closely associated if not identical with the endogenous protein substrate of the endogenous, membrane-associated cyclic AMP-dependent protein kinase.

REFERENCES

(1) C. E. Guthrow et al, J. Biol. Chem., in press.
(2) D. J. Brunswick and B. S. Cooperman, Proc. Natl. Acad. Sci., U.S.A., 68, 1801, 1971.

ACETYLATION OF RAT TESTIS NUCLEAR PROTEINS

S.R. Grimes, J.L. Irvin, and C.B. Chae, Department of Biochemistry, University of North Carolina School of Medicine, Chapel Hill, North Carolina

The in vivo incorporation of ^3H amino acids into proteins from rat testis seminiferous tubules could be blocked more than 90% by injecting cycloheximide into rat testes. During this protein synthesis inhibition, both the histones and nonhistones could be acetylated by injecting ^{14}C sodium acetate into the testes. On the other hand when either rat testis or rat liver chromatin was incubated with ^{14}C acetyl CoA, acetate was incorporated only into the histones. In either the in vivo or in vitro system acetate could be detected only in the arginine-rich histones F_3 and F_{2a1}. However it is possible that other rat testis histones are acetylated (1).

Rat testis histones have been partially characterized by chemical fractionation (2) and polyacrylamide gel electrophoresis (3). At least three histones, which were absent or present in only small amounts in rat liver, were present in rat testis in addition to the five usual fractions. The first (X_1) migrated on gels slower than histone F_1, the second (X_2) migrated between F_1 and F_3 but faster than F_1o, and the third (X_3) migrated between F_3 and F_{2b}. Preliminary evidence suggests that X_1, X_2, and X_3 are acetylated or phosphorylated forms of F_1, F_3, and F_{2b}, but present evidence is insufficient to establish or refute this suggestion.

REFERENCES

(1) E.P.M. Candido and G.H. Dixon, J. Biol. Chem. 247 (1972) 3868.
(2) E.W. Johns, Biochem. J. 92 (1964) 55.
(3) S. Panyim and R. Chalkley, Arch. Biochem. Biophys. 130 (1969) 337.